SpringerBriefs in Applied Sciences and Technology

PoliMI SpringerBriefs

More information about this subseries at http://www.springer.com/series/11159
http://www.polimi.it

Edoardo Favari · Franca Cantoni
Editors

Megaproject Management

A Multidisciplinary Approach to Embrace
Complexity and Sustainability

POLITECNICO
MILANO 1863

Editors
Edoardo Favari
PoliPiacenza
Piacenza, Italy

Franca Cantoni
DISES
Catholic University of the Sacred Heart
Piacenza, Italy

ISSN 2191-530X   ISSN 2191-5318   (electronic)
SpringerBriefs in Applied Sciences and Technology
ISSN 2282-2577   ISSN 2282-2585   (electronic)
PoliMI SpringerBriefs
ISBN 978-3-030-39353-3   ISBN 978-3-030-39354-0   (eBook)
https://doi.org/10.1007/978-3-030-39354-0

This Springer imprint is published by the registered company Springer Nature Switzerland AG
The registered company address is: Gewerbestrasse 11, 6330 Cham, Switzerland

Foreword

Megaprojects such as large transport projects, power systems, and water management systems that supply, recycle, and/or cope with rising sea levels, integrated city (re)developments, or event-driven undertakings such as the Olympics or EXPOs define our modern human landscape, at times excite the imagination, often disappoint, and all too often are the subject of outrage.

This volume presents a range of perspectives on the complexity and sustainability of megaprojects in general as well as several chapters that begin to outline what might be considered an Italian model of MPs. This multidisciplinary approach is welcome and promising, since megaprojects themselves are multifaceted and intertwined. They are at the same time physical–technical-engineering constructions, financial–economic undertakings, and sociopolitical negotiations and outcomes. Success typically requires mastery in each of these domains, and failure can come from anyone.

In line with its goal of setting up a debate among different disciplines, this volume does not come to a final definitive conclusion, but it does provide a number of important insights.

The volume begins with a set of essays focusing on the social and political aspects of projects. Barabaschi reviews stakeholder theory as applied to megaprojects and draws the important distinction between "management of stakeholders versus management for stakeholders". Cantoni and Pagnone argue that benefit shortfalls are often more important than Flyvbjerg's Iron Law of megaproject performance —"over cost, over schedule over and over again"—and that the failure to deliver what should have been possible given the scope and resources involved lies in the inadequate engagement of external stakeholders. Favari suggests that tracking benefits through a 3P lens—economic, environmental, and social—will lead to better outcomes. As the authors note in their overview:

> Applying a traditional Plan-Do-Check-Act approach simply is not enough: organizations performing Megaprojects need good managers able to apply traditional management at a tactical level, but what is missing in this approach is the long-run strategy enabling a megaproject to shape the environment where is operating in the best possible way.

The volume then turns to financial and economic conditions. Plantoni and Timpano argue that the underperformance of projects does reflect an optimism bias, but also a lack of emphasis on ex-post evaluation including the benefits delivered. Locatelli et al demonstrate that the analytical frames typically applied to shape and select megaprojects—deterministic "once and for all" discounted cash flow models—often result in project designs that are overoptimistic and "too mega", and that a real options approach, which values flexibility during project design and implementation, leads to more realistic estimates and the selection of smaller modular project designs.

The volume concludes with three pieces focused on Italian projects and practices. Maja reviews the experiences of the technically successful but economically failed Bergamo Express where the projected demand did not materialize, though the analysis does not determine whether the project failed to capture the potential demand or if the demand simply was not there. Protasoni reviews the redevelopment of EXPO15 for urban regeneration and suggests that in this instance, at least, the project was successful in balancing economic, social, and environmental concerns through a process that engaged the community in its design, and that proceeded from a very high-level allocation of open vs. developed space into specific designs. The conclusion to this chapter would be a fitting conclusion to the whole volume. It attributes the project's success to the ability to:

- "read the phenomena at different scales of interactions (from the geographical one of territory to the focused one of architecture);
- to make a synthesis of multiple knowledge (among natural and earth sciences, social and economic disciplines);
- to test decision-making analytical methodologies integrating the singular inputs of the various actors involved in the processes".

Zecchin concludes by arguing that the legal/contractual basis for projects in Italy allows for a balance between public and private interests and for pragmatic management of variations and risk.

In all, the contributions provide a useful review of the multiple perspectives required to understand, design, and manage megaprojects and serve to highlight the often underemphasized social and political aspects and the deep stakeholder engagement needed at all stages.

Donald Lessard
Professor Emeritus, MIT Sloan
School of Management
Cambridge, MA, USA

Preface

More is different

(P. W. Anderson, 1972)

Megaprojects have characterized all civilizations since the dawn of mankind. In the last decades, they are spreading not only in developed and rich countries, but almost anywhere in the world. In addition, their dimension is growing in terms of cost, impact, and complexity. Research on megaprojects is pretty recent: the first broad study has been carried out by Miller and Lessard (2000), having as a core concept that megaprojects are shaped, not chosen or planned. The message of Miller and Lessard is still valid: makes no sense to apply a planning or selection process for a megaproject, as it will not become a piece of the environment where it will be placed, but it will literally make the environment, dramatically modify the environment where it is located. Miller and Lessard focused their attention on large engineering projects (LEPs); today, researches expanded the perimeter of such studies to megaprojects, including large-scale events such as Olympics and large IT projects. The three major contributors to current megaproject research are Bent Flyvbjerg, Edward Merrow, and Peter E. D. Love. Determining whether a project is a megaproject is not immediate: there is a threshold related to the budget invested (some authors set a lower limit to 100 M€/$, some others at 1B€/$); in addition, it is pretty clear that phenomena characterizing megaprojects are emerging even in project investing an order of magnitude less, but facing a large variety of stakeholders: this is the case of several IT projects. It is pretty clear that it is a matter of complexity in the project's (internal and external) environment, but it stays unclear how to assess the level of complexity first, and how to govern it, later. What is evident is that traditional management is not working for a megaproject. More simply, traditional management is not enough. Applying a traditional "Deming" approach Plan-Do-Check-Act simply is not enough: organizations performing megaprojects still need good managers able to apply traditional management at tactic level, but what is missing in this approach is the long-run strategy enabling a megaproject to shape the environment where is operating in the best possible way.

At the same time, the public debate on megaprojects is far from being agreed upon. In the last few years, the topic is particularly relevant at the EU level. In fact the decision-making process on which infrastructure (LEP) should be funded and which not by the public sector has been criticized by new "sovranist" political parties. For example, studies both in favour and against the funding of the same infrastructure have been published by different academics, even belonging to the same institution.

To tackle this situation, in 2018 a group of researcher belonging to different institutions and academic sectors established the Megaproject Research Interdisciplinary Team (MeRIT). The group aims at:

- spreading awareness about megaprojects implementation towards public opinion;
- promoting academic research on megaprojects involving a wide range of research fields, such as engineering, architecture, management, economics, sociology, laws, and political sciences;
- being the reference point for Megaproject Stakeholders debate, by promoting discussion events involving the whole supply chain of stakeholders responsible for megaproject decision and implementation;
- Help spreading awareness about megaprojects implementation towards public opinion;
- supporting megaprojects planning, implementation, and management, transferring the findings from the academic research and the stakeholders debate to operators on the field.

The position of the MeRIT is that the research on megaprojects cannot be unique and linear: there is an epistemological limit in such an approach. Complexity requires a multiple approach in a kind of hermeneutic circle of megaprojects: the understanding of a megaproject as a whole is established by reference to the specific parts and the understanding of each specific part by reference to the whole. The circular method of hermeneutics appears the only applicable to megaprojects. No final destination and no final synthesis are achievable.

This book represents this approach. If the reader goes through all the chapters, he/she will see that independently from the reader's culture and will see the relevance of all the points represented by the authors. Chapters of this book are not intentionally harmonized, and humanistic topics are not separated from the technical ones. This way of reading and interpreting megaprojects through reciprocal contamination between different disciplines and perspectives represents and identifies the MeRIT approach.

The volume consists of eight chapters written by the constituent members of the MeRIT: jurists, economists, sociologists, corporatists, architects, and engineers. Each chapter is the fruit of reflections and studies conducted in accordance with the specific discipline of reference.

Intended for readers with different profiles and research and study interests, the book offers different reading paths, although a sequential reading of the chapters is suggested. However, other routes are advisable depending on specific needs or curiosity.

In Chap. 1, Cantoni and Pagnone present "Megaprojects. A Special Eye on Sustainability to Overcome the "Iron Law"". Unfortunately, "Over Budget, Over Time, Over and Over Again" is the summary of what is best known as Flyvbjerg's "Iron Law of Megaprojects" which theorizes that megaprojects have a history of failures in terms of missed achievement. Considering that risks tend to repeat themselves, easy-to-implement measures—some of which conventional, others less—are recommended.

Chapter 2 is about the concern of the author, E. Favari, about making sustainability effectively working in project and megaproject management: "Sustainability in (Mega)Project Management—A Business Case for Project Sustainability".

Chapter 3 "Management for Stakeholders Approach for a Socially Sustainable Governance of Megaprojects" by Barabaschi represents the critical aspect of stakeholders management in ordinary projects and, more critically, in megaprojects, providing the most up-to-date methodologies to interface and engage stakeholders at the best.

In Chap. 4, Platoni and Timpano present a rigorous dissertation about "The Economics of Mega-projects", presenting quantitative methods for decisions making and ex-post assessment of megaprojects.

An outstanding paper is presented in Chap. 5: "Using Real Options to Value Two Key Merits of Small Modular Reactors" by Locatelli, Pecoraro, Meroni, and Mancini, proposing a quantitative approach to deal with the complexity of energy megaproject according to the time to market and the existing portfolio.

Chapter 6 "Large Transportation Project: Lessons Learned from Italian Case Studies" written by Maja presents the point of view of transportation engineering on megaprojects, presenting a specific case study and providing general advice on transportation megaprojects.

Chapter 7 "Milano Expo 2015 Regeneration Process: Between Legacy and Megaproject", written by Protasoni and Roda, deals with the importance of space and space management in megaprojects, and in large-scale events in particular, providing a final Decalogue for the best implementation of them.

Zecchin in Chap. 8 presents "Megaprojects Contracts: Between International Practices and Italian Law": the legal perspective on megaprojects is often neglected, even if it is crucial for the success of such initiatives.

There is no conclusion, or final synthesis in this book, just to live up to the stated approach: centrality is the interdisciplinary debate between different disciplines, and the hermeneutic circle that cannot come, by its nature, to a final definitive conclusion.

Piacenza, Italy   Franca Cantoni
                                                                                               Edoardo Favari

Contents

Chapter 1
Megaprojects. A Special Eye on Sustainability to Overcome the "Iron Law"

Franca Cantoni and Francesca Pagnone

Abstract Over the last century, megaprojects have proven to result unsuccessfully in addressing the expected outcomes. When it comes to respecting budget and schedule constraints, megaprojects are affected by several variables and behaviors, which affect the performances. Benefit shortfalls are another perspective of underperformance, given the impact that these endeavors have on communities and groups of interest. From literature review and empirical contributions, it is possible to map and describe some managerial guidelines aimed at improving megaprojects' performances. In particular, it is interesting to look at these guidelines through the lenses of sustainability, by considering how managerial approaches should be applied and executed considering the concepts of social, environmental and economic sustainability.

Keywords Megaprojects · Complexity · Sustainability

1.1 "Much More Than Expected": Predictable and Not-So-Easy-Predictable Issues

PMBOK [1] suggests that a project is typically defined as successful when it meets the targeted scope, within the targeted timeframe and budget. Unfortunately, "Over Budget, Over Time, Over and Over Again" is the summary of what is best known as Flyvbjerg's "Iron Law of Megaprojects" [2], which theorizes that the more megaprojects turn out to be unsuccessful, the more they seem to grow their popularity around the world, attracting more and more investments every year. Megaprojects do have a history of failures, in terms of missed achievement of the planned objectives. Flyvbjerg [2], in fact, claimed that, on average, one project out of ten respects the planned budget when it is delivered, one out of ten projects is delivered on schedule, and one out of ten projects is delivered performing the expected benefits. With a simple

F. Cantoni (✉) · F. Pagnone
Università Cattolica del Sacro Cuore—Sede di Piacenza, Via Emilia Parmense 82, 29122 Piacenza, PC, Italy
e-mail: franca.cantoni@unicatt.it

calculation of probabilities, it is possible to say that, out of a thousand projects, only a single one is delivered with respect of the three main deliverables/constraints of the Triangle.

After having underlined the substantial differences between projects and megaprojects it is natural to ask whether these parameters (time, budget and scope) can be considered as exhaustive also for megaprojects. [3] specifically mentions the impact that these endeavors have towards their stakeholders and society: not only megaprojects have a wide range of stakeholders that are directly impacted, but also communities and societies which are supposed to benefit from them. This can lead to still consider megaprojects' success basing on the time of delivery, budget—since megaprojects are often sustained and funded by taxpayers' money—and scope delivery, meaning the value perceived by society.

Value creation can be considered under two main perspectives; firstly, the outcome-based one [4] considers the value a megaproject creates right after its completion: typically, it is related to the financial revenues produced for project sponsors and labor force which contributed to its creation. Instead, the system lifecycle-based perspective includes tangible and intangible values developing through the "operations phase" (i.e. when the project is actually performing its tasks) [3]. Basing on these concepts, internal stakeholders of megaprojects can be more easily associated with the outcome-based perspective of value, i.e. the value drawn from project management practices deployment in order to reach specific and pre-determined goals. Instead, external stakeholders seem to be more associated to the system lifecycle-based perspective of value, meaning that the positive impact must be perceived on the outcomes rather than the project's metrics.

1.1.1 Cost Overruns

According to several analyses and investigations conducted by the most relevant authors of the field [5, 2, 4], cost overruns are a major feature and challenge of megaprojects, which prove to be empirically more expensive than expected. The problem of cost overrun is, therefore, not much of a potential cause of underperforming megaprojects, but an actual recurring event, which gives a great hint for further analysis. Furthermore, it is an occurrence that can be related to not just one field, but several areas of economic development in which megaprojects have an important role: from infrastructures, to energy production and extraction, to information technology. There is also not a pattern that concerns only one country, so the phenomenon can be considered with an "internationalised" exposure. Finally, what seems to be the most striking aspect of these conclusions is the timeframe: the results show that, despite the passing of time and the improvement of technologies and knowledge worldwide, costs overrun have not been mitigated.

What is the fault for these continuous cost overruns? Can it be possible to mainly blame contingencies, here intended as external and unexpected events having an impact on the forecasted costs? Indeed, there is an extent of events that are uncontrollable by megaproject managers. Budget shortcomings are the most discussed topic concerning megaprojects' under performances. In fact, there is an interesting asset of opinions coming from scholars and researchers on why budgets (or forecasts) turn out to be inconsistent so often. For instance, Flyvbjerg et al. [6] consider four main areas of causes that might explain the cost underestimation trends: these include technical, economic, psychological and political reasons. Indeed, what appears as key feature of Flyvbjerg's point of view on cost underestimation is, indeed, the psychological perspective of this phenomenon. In fact, he gives great importance to the so-called "optimism bias" phenomenon, for which project promoters do underestimate costs owing to the "cognitive predisposition […] to judge future events in a more positive light than the actual experience" (pp 5–15). Indeed, besides the phenomenon of optimism bias, he definitely leaves room for the "strategic misinterpretation", i.e. the underestimation of costs made on purpose. However, he keeps the main cause for cost underestimation on the dual perspective of deceptive (i.e. strategic misinterpretation) and non-deceptive (i.e. optimism bias) practices.

The lack of capabilities for psychological and political reasons to entirely justify cost overruns is also validated by the perspective of Love et al. [7]; according to this perspective, these two explanations do have some kind of adherence to reality, but are simply not enough to explain these practices as a whole. Instead, Love focuses its investigation in the causes that come from the independent factors that provoke cost overruns, and their interdependencies with this phenomenon. Therefore, the point expressed by Love and the school of thought of the "evolution theorists" [8] reminds that the investigation must not just be done on the deceiving actions in the "FEL" (i.e. Front End Loading) phase of megaprojects, but also in the development and implementation phases, which can bring external and unforeseen events to potentially change the scope of a megaproject and, as a consequence, its budget and its schedule.

1.1.2   Schedule Delays

Among the three main performance metrics and shortfalls of megaprojects, schedule delays can be considered as the "most consequential" variable; as seen from the previous examples of the historical overview, often schedule delays are a direct consequence of a not well-defined strategic vision for a megaproject, brought about by significant misalignment of stakeholders. This, for instance, happened with the Three Gorges Dam megaproject, which was significantly delayed with respect to the time of its initial conception, owing to political disagreements and different interests in conflict.

In the same way, schedule delays are a direct consequence of the situations of cost overruns, considering the time needed for megaproject promoters to source for new financial sponsorship, whenever the megaproject finds itself in the "breaking" phase of the traditional Break-Fix Model. One of the most striking example of these situations is the Channel Tunnel megaproject, which, at some point in its construction, opened the doors to public share offerings in order to have more financial availability.

Besides being a consequence of other shortfalls, megaprojects' delays can be considered as triggers for some other risks and failures, such as the growth of interest and exchanging rates of financial resources, if the project takes longer than expected [9]. Moreover, schedule delays can have consequences in terms of adaptation to the market; in fact, if a project is in the public eye since its conception and planning phases, this definitely has an impact on the market where the project operates and its competitors. Other market actors could be triggered by the news of a megaproject and improve their own competitiveness. Indeed, if this situation is applied to a megaproject with schedule delays, the project will end up being completed when competitors will already have made their moves. Overall, the delayed megaproject will turn out to have weakened competitiveness, compared to its market.

1.1.3 Benefit Shorfalls

As well as the misleading forecasts in terms of costs and budget, also misleading forecasts in terms of future demand can seriously undermine the success of a megaproject [9] since, in some extent, the existence of a certain demand from society is the major reason why megaprojects are conceptualized in the first place.

Flyvbjerg gives space for the use of wrong methodologies as a reason for mismatching forecasts; hence, it is possible to consider it among the causes of poor forecast reliability, especially when thinking about the fact that, sometimes, change management in organizations towards new technological tools can be challenging. Moreover, particularly in the field of transportation, the market has lived decades of continuous development of complexity, with new market entries and increased competition. As for other identified reasons for misleading forecasts, Flyvbjerg considers the potential lack of data available, which could be further developed by conducting so-called "stated preference analysis", investigating on consumers' choices and behaviours. Furthermore, the forecasts can be undermined by discontinuous behaviours of the analysed individuals, which might be led to change owing to the change in complementary factors, i.e. factors mostly related to the "customer experience", but not directly considered in the demand forecast since they are not directly related to the core service offered. Exogenous factors also play a significant role; exogenous factors are intended as the unexpected and uncontrollable events of social or macroeconomic nature, which indirectly affect the actual demand and might widen the gap with the forecasts. On this wave, also political decisions can relevantly affect the demand for a megaproject; for instance, if the demand is forecast according to certain regulations, which are then not legalized, enforced and/or enacted, there is certainly room for some demand to be affected by these changes.

A parallel reflection needs to be done also on the phenomena of "appraisal bias"; this is particularly related to the forecasts made by consultants and project promoters, who might tend to produce projections that produce more plausible or desirable outcomes. Obviously, both consultants and project promoters have motives to be "optimistic", especially when forecasts are created in order for projects to gain financial sponsorship or regulatory permissions. This concept of optimistic views on forecasts seems to significantly resemble the one related to cost projections. Again, also in this perspective, Flyvbjerg does not push the investigation too far, defining these mismatches as "implicit" and not specifically referring to any intentional actions coming, above all, from project promoters. However, it seems quite easy to associate these demand "exaggerations" with intentional actions coming from project promoters, especially in the phase in which projects seek for financial sponsoring and/or regulatory approval. In conclusion, it is safe to say that misleading demand forecasts are not just the results of statistically random casualties, but there are concrete bases to believe that forecasts are often biased by the aforementioned factors.

1.1.4 The Influence of Secondary Stakeholders

Among the not-so-easily-predictable issues megaprojects can result as unpopular and receive the opposition of local communities, as they frequently perceive them as environmentally disruptive or with general negative impact on society's well-being. Gellert and Lynch [10] express a very harsh opinion on megaprojects, claiming their tendency to be disruptive towards the environments and the local communities. What clearly emerges from a literature review on this topic is that the failure of megaprojects is particularly analysed under the spectrum of the so-called "indirect stakeholders", i.e. society, environment and institutions. This is definitely caused by the fact that the broader impact is one of the key features that differentiate megaprojects from traditional projects, and has, therefore, to be considered in this optic.

It is clear that, under all circumstances, megaprojects should have the overall goal of accomplishing the objectives of their stakeholders, in which institutions are definitely included. On this concept, Di Maddaloni and Davis [7] also aligned by claiming that the objectives that are met in terms of costs and schedule are never enough to define a megaproject as successful: there have to be real benefits that have to be set in advance, and not just assessed after the project is implemented and considered as some sort of "collateral effects".

A key component of the challenges for megaprojects to actually bring results on the institutional perspective is the concept of considering stakeholders as the ones who provide with resources [11]; it is legitimate to wonder whether this idea is consistent or not for megaprojects. Once again, the answer can be found in the very own definition and key characteristics of megaprojects, which clearly entail the involvement of wider classes of stakeholder; therefore, not just the ones who bring physical resources should be considered. This idea is further validated by the fact that one of the main causes for benefit shortfalls is brought by a lack of public participation [9]. This idea is also expressed in the work of [12] in which a statistical analysis on 44 megaprojects with the aim of identifying success factors highlighted that one of the major critical factors for a positive outcome delivery is local public acceptability of megaprojects, which affects them positively, together with the lack of litigation procedures and the multiculturality of project teams.

On a different perspective, benefit shortfalls can also be seen not just as the poor consideration of some classes of stakeholders, but also as a lack of integration and alignment of interests for the considered ones. Biesenthal et al. [13], for instance, gather the attention on megaprojects having governance structures that turn out to be polycentric; this is mainly caused by the complexity of the stakeholders' array, as already mentioned, but also caused by the fact that several stakeholders have key roles in megaprojects' deliveries. As an example, we can consider politicians, who can give media exposure and public approval with their influence. At the same time, regulators and regulations can give or deny the green light for a megaproject on a concession standpoint. In addition, financing sources are equally essential resources, being the size of megaprojects so considerable. Therefore, megaprojects' stakeholders and actors are often leaders or senior positions of their organizations, used to being at the top of their own governance systems. With these premises, the consequence is a challenging formation of a governance structure that is the megaproject's own, and hierarchical decision making is harder to develop, especially since direct stakeholders all bring essential resources to the table.

Indeed, each stakeholder group comes with different and, sometimes, contrasting interests and desired outcomes. The references found in classic megaproject literature [14] describe stakeholders' destruction of megaproject's value when considering their own interests and desirable outcomes as the megaproject's drivers for success. This happens when the project's overall strategic vision is not aligned with stakeholders' objectives and/or stakeholders are vaguely aware that there is a strategic vision that aligns all the expected outcomes [15].

1.1.5 Corruption in Megaprojects

In between the investigation on causes for time, scope and budget under performances, it seems essential to put a significant stress on the phenomenon of corruption, which has surprisingly found few literature at support. One of the most interesting works on the topic has been elaborated by, in an article that puts some light on the practice and finally includes them in the causes for shortfalls in megaprojects' delivery. For what concerns large projects, the article makes an interesting insight on how corruption can more easily gain space in contexts where megaprojects are developed and implemented. This is due to megaprojects' key characteristics, such as the size. In fact, where there are complex networks of stakeholders and many individuals involved, it is easier for some actors to enact briberies and illegal practices, and hide them, given the whole complexity of the system. More contracts and more parties collaborating generate more opportunities for bribery and power abuse, which is also encouraged by the relevant presence of public institutions participating and having access to potentially manipulable offices, regulations and control procedures. In addition, since megaprojects that involve public institutions are more than often in the spotlight of public opinion and mass media, there is a significant tendency in attempting to covering sensitive information that could be used and misinterpreted by the media and, as a consequence, by society.

An important take-home is related to the impact of corruption on megaprojects, which has an extent on both megaproject and project success. Firstly, projects are impacted in terms of costs, as the infrastructural ones tend to increases and have significant additions. At the same time, there is a tendency for delay, brought by initial infrastructures that prove to be unsuccessful. In conclusion, what really occurs is a "sub-optimal allocation" of resources in order to privilege private gains and interest of some parties involved. Interests that are clearly not included in the ones that are clearly attributable to stakeholders being stakeholders, but go beyond the coordination and alignment of actions that should naturally be performed in order to deliver the expected value from the implementation of megaprojects.

1.2 Overcoming the Issues

Indeed, a winning strategy would entail the utilization of a rigorous method of cost forecasting, especially concerning the moment in which the forecast is made. As previously discussed, misinformation and underestimation of costs is also caused by different forecasts made in different stages of development. This creates confusion and does not allow to have a clear idea on what going over budget really means. Locatelli et al. [16] do express a solution to this problem; the clear setting of a stage of development for each project, in which the official and most reliable forecast is made. Indeed, the most indicated stage of development varies depending on the specifics of the project, as well as the sector of reference.

Flyvbjerg describes with detail what is known as the "conventional" megaproject management approach; based on various empirical examples, such as the Great Belt bridge in Denmark, the author gives an overview on how megaprojects are typically managed, highlighting the central role of the government and state-owned enterprises. Among the various identified characteristics, the lack of attention to the external effects of the project and the negatively affected stakeholders is major. In addition, Flyvbjerg also recognizes a lack of in-depth analysis especially concerning risks. This underestimated stakeholder involvement in the traditional megaproject approach is considered as the main trigger for consequent public dissatisfaction.

The last century has seen a growing tendency of private organizations being involved in the financing of megaprojects; as analysed and described by traditional literature, some aspects have to be considered on the entering of private entities in megaprojects' development: for instance, it is necessary to consider the added value that private entities may bring in terms of technical knowledge and capabilities is construction and cost controlling. It may also have positive outcomes in terms of risk identification, owing to the different perspectives that the private sector might bring to the planning and risk assessment phases. When private and public sector collaborate in this context, there is definitely a need for long-term commitments on both sides, which enforces a formal relationship between two or more entities and implies more specific inputs and outputs coming from all the actors participating.

The more accentuated involvement of the public and the indirect stakeholders is considered as absolutely necessary in order to overcome the traditional megaproject management approach, which leaves most of the roles and responsibilities to the government. If a management approach that is more focused on the various stakeholders is applied, stakeholder groups and society representative should be involved in the development process and informed with transparency, right from the earliest stages of development. Firstly, this process can more efficiently happen if there is a sharing of clear information through public means of communication towards the public. This action would be according with the principle of transparency, which governments involved in megaprojects should maintain. At the same time, the clear flux of information must also be collected from the stakeholders, with tools such as surveys, public hearings, committees etc.

Overall, it seems that the most problematic aspect that challenges megaproject management is the involvement and alignment of many stakeholders, as it was initially assessed by identifying megaprojects' complexity. For this reason, what megaproject management might really need is a more rigorous and sensitive method to involve and coordinate multiple stakeholders in order to ensure the alignment of their interests, their effective contribution with resources and their assessment of realistic results. On the topic, literature [14] has theorized interesting solutions and guidelines, which reconnect adequate project management practices to the very own creation of value coming from megaprojects; this value is, in fact, generated from all the joint activities of megaproject actors, organized in organizational platforms. These platforms are described as organizational structures that store multiple actors and their resources, whether they are tangible or intangible. Organizational platforms, in traditional business processes, involve actors from both the production and the consumption. There is, therefore, plenty of different actors involved, just like it occurs for megaprojects. For this reason, literature often describes megaprojects as "temporary organization" [17].

Hence, the main procedures that take the single actors and transform them into an organizational platform for a megaproject's development, is the matching of the actors with their inputs and resources, and the identification of an overall system-level goal that aligns and unifies the goals of each single actor. By doing so, the system-level goal becomes the key to a megaproject with a clear, strategic vision, which, as claimed by literature [15] as one of the main success factors. The very own interaction among actors paves the way for an alignment of resources and interests with the creation of interorganizational coordinating bodies, emerged from the development of joint activities. By interaction, actors are able to identify all the resources, including the most critical ones, and the various responsibilities that can, then, determine calibrated ownership and decision-making.

Interorganizational coordinating bodies have the functionality of facilitating the joint activities and enhance structured governance systems, which are key drivers for megaprojects' success coordination creates a key role in collecting stakeholders' inputs and expected outcomes, in order to define a clear strategic vision. Basically, coordinating actors identify what value stakeholders are expected to capture, what resources they can bring to create this value, and what resources represent the most critical ones. Hence, there is alignment of these various perceptions of value, and alignment of the expectations of single actors. At this stage, the process of alignment is crucial for the identification of project stakeholders as one unique organizational platform: it is also the most challenging aspect of stakeholders' management, as the strategic system-based goal that emerges from the alignment of all the stakeholders' goals still has to be able to represent these stakeholders. In other words, stakeholders should be able to still see what is in it for them when the organizational platform is formed and, most importantly, stakeholders have to be able to logically recognize what perspectives of value they will not be able to reach and why.

Interorganizational coordinating bodies have, therefore, this very complicated task to accomplish. In order to do so, they can apply some approaches and tools to more efficiently create the ideal context for stakeholders to properly collaborate. Firstly, the identification of a strategic goal and the concrete reachable value of a megaproject, coordinating bodies have to perform an in-depth analysis on the expressed stakeholders' objectives; this enables to find actions and practices that will lead to a kind of "intersectional" result, which may be the right outcome to address the expectations of more than one stakeholder, if not most of them. To express it in geometrical terms, this phase of the process resembles the search of an "intersection" among expected outcomes. Once this intersection is found and the alignment among objectives is performed, coordinators can and have to use the means of transparency to clarify the outcomes with stakeholders as much as possible. This will prevent misalignment on the expected outcomes, but will also lead actors to realize that, even with different extents, the value they wish to capture from the megaproject is considered in the "bigger picture"; this would be especially beneficial for secondary stakeholders and their participation in the megaproject's development. More clear objectives communicated to all the involved parties are definitely a tool that coordinators can use to minimize the cultural gaps and enhance effective collaboration, which are two of the main and most occurred challenges [18].

Furthermore, another aspect has to be considered in terms of coordination among stakeholder; promoters and managers do also have to implement leadership practices. Not only taking advantage of leadership allows a better performing decision-making, but also allows to better track progress on a greater perspective. Leadership in megaprojects is seen [15] as an essential resource to be added to managers of daily operations, since megaprojects do also need an intense "institutional management", more intensively but together with "technical management" [19].

On the topic of coordinating and interorganizational bodies, it is interesting to consider the research done on the so-called "Special Purpose Entities" (SPEs), particularly relevant in the context of megaprojects, but, according to some literary examples, significantly under-investigated. SPEs are commonly defined as fenced organizations with pre-defined purposes and their own legal identity. Typically, the organizations that act as SPEs in project management are the ones who collaborate in project financing and project partnering. If employed for project financing purposes, SPEs contribute in assessing and isolating risks for potential investors; they act as a further proxy for external financiers. For what concerns their project partnering usefulness, SPEs are the legal entities that allow partnering among stakeholders, in order for their interests to be aligned. Basically, project stakeholders, including private and public entities, can use SPEs to create partnerships within each other or form actual joint ventures; they, therefore, create, a separate organization made up of organizations. Megaproject SPEs are typically physical organizations, with a defined timeline and a define purpose; they are typically set in a venue that is close to the location where the megaproject is constructed, and have their own legal identity, which allows them to be entitled for assets, liabilities, people employment et cetera, just like any other organizations. Moreover, another interesting feature is that, being SPEs "fenced" entities, if their shareholders go bankrupt, the SPE's assets are not liable to recover from this bankrupt; therefore, to the eyes of financial lenders, SPEs are definitely a reliable instrument. Overall, it is possible to consider SPEs as real organizational platform with legal identities. Hence, their existence and proper employment can resemble the creation of an interorganizational coordinating bodies, accomplishing the objectives of alignment, coordination and co-evolution that these bodies are aimed at. Furthermore, given their fenced nature and independent identity from the forming organizations, they can also be perceived as risk regulators and reliable means of collecting financial resources. In other words, SPEs are able to create the governance structures that companies implement and that megaprojects, as temporary organizations, would need.

1.3   Conclusions: Sustainability as a Key of Interpretation

Many are the inputs that can be collected by literature research and empirical examples. In order to provide with a main theory that unifies all the guidelines offered by literature, it would be useful to introduce the concept of sustainability. The OECD [20] defines it as "The continuation of benefits from a development intervention after major development assistance has been completed. The probability of continued long-term benefits. The resilience to risk of the net benefit flows over time." Considering the concepts of benefits and resilience that the definition of sustainability comprehends, it is possible to interpret sustainable behavior as a key to solve and/or mitigate criticalities in megaproject management, as well as to solve criticalities and challenges in general world economic, social and environmental development. Guidelines and hints related to megaprojects can be mapped into the three main sustainability pillars; economic, environmental and social sustainability.

1.3.1   The Economic Perspective

In terms of economic benefits that sustainable megaprojects should bring, it is necessary to consider fair financial rewards and earnings for stakeholders that bring resources. This fair economic return should be rightfully forecasted, together with the projections of public demand of a megaproject's services. As previously seen, literature on megaprojects offers hints and insights on how to perform these activities fairly and lawfully, with consequent benefits to the overall performance of a megaproject. More precise forecasts with less "optimistic" approaches, the decision of assessing forecasts that are always made in the same phases of the projects, more investigation on the data related to similar megaprojects' costs, as well as precautionary practices of identification and assessment of risks: these are all guidelines that can be drawn by literature and empirical evidence, and all can lead to a more sustainable economic development. In this sense, literature also gives a perspective for further reflection on how stakeholders should collaborate towards a clear definition of what are the expected economic benefits for them. This guideline would be an attempt to overcome situations in which corruption episodes occur; needless to say, the effectiveness of this guideline to reach the afore-mentioned result can be doubted and further investigated; for the level of this analysis, this guideline can be considered as a starting point for mitigating the challenge of corruption.

1.3.2 The Environmental Perspective

Modern society is further and further supporting governments and regulations promoting environmentally sustainable business practices; environmental consciousness should, therefore, be of incredible concern for megaprojects promoters and executors [1]. What is more useful to say on this topic is that, according to the literature and empirical examples, nowadays contemporary organizations do have the tools and frameworks to implement environmental sustainability into megaprojects' development. For instance, the several groups of interest operating on national and international levels can, in this sense, provide organizations with support on feasibility studies that also consider environmental impacts of megaprojects. Since several organizations already have focused their efforts in operating following sustainable environmental practices, there already are positive premises for megaprojects' sponsoring companies to translate their knowledge on environmental sustainability into a more conscious planning and execution of megaproject with potentially massive environmental impacts [18].

1.3.3 The Social Perspective

This sustainability pillar has a major role in determining megaproject's performances and, therefore, deserves to be treated as a conclusive topic and main insight for further research. Basing on the findings of this essay, social sustainability is an essential key to overcome all the megaproject challenges that have to do with public audiences and community members, who should be the very first perceivers of benefits.

Not only, social sustainability can be considered as the key interpretation to a more effective stakeholder management, applying all the practices described by literature and empirical examples, aimed at identifying stakeholders' interest, resources, outcomes and potential ownership and decision-making power. Stakeholders, at all levels, should be included at all stages of development; they should be given the necessary information and, at the same time, the possibility to provide each other all the necessary information. Their goals should be aligned with each other and translated into one system-level goal, which becomes the strategic vision of the megaproject. In order to perform these activities of integration, mediating and coordinating entities, such as the SPEs acting as interorganizational bodies should be assessed.

In the same way, empirical evidence paves the way to interpreting alignment also on the internal perspective, regarding the attention that must be made on the "institutional capabilities" that professionals involved in megaprojects should embody. In order to work in a context where "institutional management" is more consistently valued than "technical management", the recruitment and selection of project actors has to be performed bearing this idea in mind with the expectation that a selection process that considers an asset of soft skills leads to a project management practice that has an institutional connotation, which is an essential trait of megaproject management. Among stakeholders, as often highlighted in the previous paragraphs of this chapter, communities do have a major role to achieve social sustainability. Communities, either on a local, national or international level, must be integrated into the life of a megaproject that is planned to exist with the aim of improving the quality of life. Integrated communities can give precious hints to megaprojects promoters in order to plan, build and execute infrastructures or any other kind of endeavors that are actually responsive to their needs. The hints coming from communities have the ultimate goal of better addressing the resources that stakeholders and megaproject promoters and sponsors to a socially sustainable final product. The social sustainability will, consequently, turn into public acceptance and, thus, an increase in the demand for the offered services. Therefore, social sustainability is not just about the social impact of a megaproject per se, but it also can bring collateral benefits falling into the economic and the environmental dimensions of sustainability.

References

1. PMBOK® Guide – Sixth Edition (2017).
2. Flyvbjerg B (2017) Introduction: the iron law of megaproject management, in Bent Flyvbjerg ed The Oxford Handbook of Megaproject Management (Oxford: Oxford University Press), Chapter 1, pp 1–18
3. Artto K, Ahola T, Vartiainen V (2016) From the front end of projects to the back end of operations: managing projects for value creation throughout the system lifecycle. Int J Project Manage 34(2):258–270
4. Gellert P, Lynch B (2003) Megaprojects as displacements. Int Soc Sci J. https://doi.org/10.1111/1468-2451.5501002
5. Di Maddaloni F, Davis K (2017) The influence of local community stakeholders in megaprojects: rethinking their inclusiveness to improve project performance. Int J Project Manag 35(2017):1537–1556
6. Edkins A, Geraldi J, Morris P, Smith A (2013) Exploring the front-end of project management. Eng Proj Organ J 3:71–85. https://doi.org/10.1080/21573727.2013.775942
7. Flyvbjerg B, Bruzelius N, Rothengatter W (2003) Megaprojects and risks: an anatomy of ambition. Cambridge University Press, Cambridge
8. Ahiaga-Dagbui DD, Smith SD (2014) Rethinking construction cost overruns: cognition, learning and estimation. J Financ Manag Property Constr 19(1):38–54
9. Locatelli G (2018) Why are megaprojects, including nuclear power plants, delivered overbudget and late? reasons and remedies. Report MIT-ANP-TR-172, Center for Advanced Nuclear Energy Systems (CANES), Massachusetts Institute of Technology
10. Locatelli G, Invernizzi D, Brookes N (2017) Cost overruns—helping to define what they really mean, vol 171, Issue CE2. Institution of Civil Engineering. https://doi.org/10.1680/jcien.17.00001
11. Love PED, Ahiaga-Dagbui DD, Irani Z (2016) Cost overruns in transportation infrastructure projects: sowing the seeds for a probabilistic theory of causation. Transp Res Part A Policy Pract 92:184–194
12. Shenhar A, Holzmann V (2017) The three secrets of megaproject success: clear strategic vision, total alignment, and adapting to complexity. Proj Manag J 48(6):29–46
13. Biesenthal C, Clegg S, Mahalingam A, Sankaran S (2017) Applying institutional theories to managing megaprojects. Int J Project Manag 36(2018):43–54
14. Locatelli G, Invernizzi DC, Brookes NJ (2017) Project characteristics and performance in Europe: an empirical analysis for large transport infrastructure projects. Transp Res A Policy Pract 98:108–122
15. Miller R, Lessard DR (2001) The strategic management of large engineering projects: shaping institutions, risks, and governance. MIT Press, Cambridge
16. Locatelli G, Milos M, Kovacevic M, Brookes N, Ivanisevic I (2017) The successful delivery of megaprojects: a novel research method. Proj Manag J 48(5):78–94
17. Lundin RA, Söderholm A (2013) Temporary organizations and end states. Int J Manag Proj Bus 6(3):587–594
18. Sainati T, Brookes N, Locatelli G (2017) Special purpose entities in megaprojects: empty boxes or real companies? Proj Manag J 48(2):55–73
19. Flyvbjerg B, Skamris Holm M, Buhl S (2002) Underestimating costs in public works projects: error or lie? J Am Plann Assoc 68(3):279–295. https://doi.org/10.1080/01944360208976273
20. OECD Annual Report (2002) Public Affairs Division, Public Affairs and Communications Directorate, Published under the responsibility of the Secretary-General of the OECD

Chapter 2
Sustainability in (Mega)Project Management—A Business Case for Project Sustainability

Edoardo Favari

Abstract Sustainability in a 3P framework is becoming a key driver in megaproject feasibility assessment according to International Financial Institutions. Currently, in several cases, sustainability is still used as a fair topic to talk about by CEOs and presidents of Companies and Institutions without a concrete plan to implement it. This paper aims to explore the strategies for a real implementation of 3P sustainability in projects and megaprojects, integrating the three pillars of sustainability into a cohesive whole to make it works not only at year-end speeches to shareholders, but also to fruitfully include sustainability into the strategy of organizations.

Keywords Sustainability · Project management · Business case · Megaproject · Feasibility

2.1 Focusing on Sustainability of (Mega)Projects

Sustainability in the triple bottom line (TBL), also referred as 3P framework (People, Planet, Profit) is becoming a key driver for business strategy in the twenty-first century [1]. Today, the question "How can we develop prosperity without compromising the life of future generations?" is generally recognized worldwide as a basic need for setting strategic goals of organizations [2]. In fact, according to [3]: "As of 2016, we are consuming 1.6 planets worth of resources, and based on current trends, this will increase to two planets." In scientific literature, sustainability is a fair new topic [4]. Moreover, sustainability goals are expressed in the United Nations' "The 2030 Agenda for Sustainable Development" [5], adopted by all United Nations Member States in 2015. In this document, the "P" principles became 5: in addition to People, Planet and Prosperity (Profit), "Peace" and "Partnership" have been added. In the mentioned UN resolution, 17 goals to sustainability are expressed. The Sustainable Development Goals are:

E. Favari (✉)
PoliPiacenza, 29121 Piacenza, Italy
e-mail: edoardo.favari@polimi.it

© The Author(s), under exclusive license to Springer Nature Switzerland AG 2020
E. Favari and F. Cantoni (eds.), *Megaproject Management*,
PoliMI SpringerBriefs,
https://doi.org/10.1007/978-3-030-39354-0_2

1. No Poverty
2. Zero Hunger
3. Good Health and Well-being
4. Quality Education
5. Gender Equality
6. Clean Water and Sanitation
7. Affordable and Clean Energy
8. Decent Work and Economic Growth
9. Industry, Innovation, and Infrastructure
10. Reducing Inequality
11. Sustainable Cities and Communities
12. Responsible Consumption and Production
13. Climate Action
14. Life Below Water
15. Life on Land
16. Peace, Justice, and Strong Institutions
17. Partnerships for the Goals.

Nonetheless, in the remainder part of our paper, we will continue to refer to the 3P framework as it is more common and easy to be applied at project management level, and we will assume that an organization or a project, to be considered "sustainable", must address all the 17 goals mentioned above.

In the last decades, sustainability and corporate sustainability have become a stilted and moralistic topic, and the public opinion feels that it is used to clean the conscience or to get public consensus, without concrete argument. As a matter of fact, what is missing today is the deployment of these strategic goals related to sustainability into the operational activities of organization [6]. According to all the scientific sources consulted [3, 7–9], the discipline that can help organizations achieving their strategic goals is project management: in fact, in order to survive and prosper in a global environment that is continuously evolving, organizations must endlessly develop changes in their way of doing business, and project management is a key skill capable to execute these changes in a structured manner [7].

Currently, no Project Management frameworks (PMI and IPMA above all) include sustainability knowledge areas or specific processes [8, 10]. Moreover, no specific input nor output in project management frameworks considers sustainability in the 3P approach, but only some (rare) specific points referring to one of the three pillars can be found [7].

A project manager, alone, can't succeed in including sustainability goals in his/her project, if these goals are competing with goals set by the project sponsor and/or PMO in the project charter [11]. In order to influence the way project management is carried out, and to include sustainability into project goals, decisions must be taken during the business analysis phase (i.e. the phase during which decision on what project solution is the best to fix the problem or opportunity addressed is taken), analysed and included into the business case, approved through a formal Go/NoGo decision, so that project managers are fully empowered to develop projects in a controlled manner [12, 13]. In fact, no organization would leave to individual project managers the discretionary power to make decisions related to sustainability [14]. Indeed, sustainability decisions could cause expenses that cannot be justified simply from the project's point of view, because sustainability has long-term goals, whereas project goals are generally short-term oriented [7]. In addition, extra expenses caused by addressing sustainability are often due to the internalisation of externalities [1], which, again, can't be justified at project level without a previous approval of the project sponsor.

Decisions on sustainability must be set at a strategic level and, then, cascaded to professionals managing the tactical level, otherwise not even the best project managers having the highest competence on sustainability will have the power to implement sustainability principles [15]. This is very clear, for example, in civil (mega)projects where sustainability rating systems, such as LEED for buildings or Envision for infrastructures, can be applied (see following sections). These frameworks cause direct costs of resources to carry them on all along the project duration, and indirect cost for extra-design required and for non-standard solutions to address sustainability goals [16, 17]: these direct and indirect extra costs cannot be established in the project phase, but must be determined before the decision of developing the project is taken.

This approach for including sustainability goals into projects is potentially in conflict with the definition of project management: "The application of knowledge, skills, tools and techniques to project activities to meet the project requirement" [8]. In fact, if project requirements only include project goals and requirements needed for the project's specific purpose, probably sustainability has no chance to be present in project management. Instead, project requirements must include points that consider the organization's long term strategy: in this way, sustainability can be seen as sustainable for business purposes [18].

This fact is particularly critical on megaprojects that are often developed through Special Purpose Entities (SPEs): in fact, SPEs are established just for executing the single megaproject, and it is harder and harder to include into their goals something exceeding the megaproject itself. To correct this phenomenon, most of International Funding Institutions (IFIs), starting from the 90s, adopted specific policies related to sustainability goals that committed to be applied to any funded project [19–23]. So, in the case of projects developed by SPEs, the strategic part to be adopted into the business case and, following, the project charter, is to come from the sustainability policy of the funding entity.

This paper focus on identifying principles for developing a business case for a (mega)project including sustainability, in order to evaluate the implications of incorporating a 3P framework in the way projects are selected, tying environmental and social goals to economic and financial sustainability, incorporating both direct and indirect benefits. Sustainability goals are economically sustainable if considered in the organizational framework (long-term). If only the short-term focus is maintained, sustainability goals cannot be supported because this violates the economic sustainability. "It is no longer just about risk and compliance, but also about innovation and opportunity and how to simultaneously achieve excellence in both sustainability and financial performance. (…) not just thinking about corporate social responsibilities and risks, but also about corporate social opportunities" [24].

In the following section of this paper, the Author will list strategies to implement sustainability in projects within a 3P framework, classified within distinct business sectors, and provide hints for evaluating direct and indirect costs and benefits, to work as an input for a business case and an economic evaluation of the project [25–31].

2.2 The Business Case

According to [8]: "A business case captures the reasoning for initiating a project". For megaprojects, several evaluation approaches are possible. According to what presented in this paper, as long as 3P sustainability isn't assessed in a integrated way, in a kind of sustainability iron triangle, in which each environmental or sustainability goals is clearly linked to economic and financial ones, sustainability will stay a fair topic, only good for public speeches, but minimally addressed. Moreover, in some circles, sustainability is even addressed from a mystic perspective, and this does not support the pragmatic analysis of balancing the breakeven of implementing it.

All along this paper, we provide points for evaluating costs and benefits sustainability-related, that allows to develop detailed business analyst to carefully evaluate the feasibility of sustainability goals. The paper provides a list of direct and indirect costs and benefits to be included in business cases to consider sustainability principles. Furthermore, a description is given of which tools and techniques are needed by the project manager to lead a sustainable project: this is done to let the business analyst, when drafting the business case, being able to evaluate operational activities needed for reaching project's sustainability.

The hope of the Author is that a detailed and accurate business case is drafted for any project, to carefully evaluate the level of sustainability that is sustainable for the performing organization and the community of stakeholders in which the project take place. Only by avoiding dealing with sustainability in a rhetoric manner can lead project to concrete sustainability.

2.3 Drafting a Business Case for (Mega)Projects Sustainability

2.3.1 Economic Direct and Indirect Benefits of Applying Sustainability to (Mega)Projects

In this section we will enlist sustainability topics to be evaluated when drafting a business case for a project. These are points that can provide benefit to the performing organization and stakeholders of projects, including SPEs and funding entities.

Financial profit

- Direct cost savings due (mainly) to environmental sustainability, such as reducing the use of material and energy during project development;
- Reduced operations costs including fines and litigation costs;
- Regulatory compliance: "Government regulations ad industry codes of conduct require that companies must increasingly address to sustainability" [24]. Addressing sustainability in a project will enable the result of the project to have a longer lifecycle and to avoid extra-cost of compliance to new regulations and policies;
- Lower capital costs;
- Organization's share value increase and received investments due to implementation of sustainable practices.

Benefit related to customers and stakeholders

- Increased Customer and stakeholders satisfaction;
- Innovation related to listening to the voice of the customer/stakeholder;
- Market share increase of the performing organization due to the increasing demand of sustainable products worldwide;
- Mastering sustainability can lead to the development of new and innovative product and services;
- Stronger reputation;
- New market opportunities.

Operational benefits

- Innovation in processes;
- Productivity improvement due to operational waste minimization;
- Design for Sustainability [32];

Organizational benefits

- Employees satisfaction, personnel turn over reduction, and attractiveness for qualified professional, in implementing sustainability principles;
- Improved shareholders and stakeholders relationship;
- Lower level of risk and impact by implementing sustainable practices;
- Increased resilience and organizational learning;
- Better decision-making process, improving resilience, due to the resolution of ethical dilemmas and improving corporate governance, becoming an accountable organization;
- Performing organization's brand value increase, leading to increased sales due to better corporate reputation.

2.3.2   Direct and Indirect Cost of Applying Sustainability to (Mega)Projects

In the following sections we will enlist specific activities to be performed in order to guarantee the sustainability of a (mega)project. We can't make a list of direct design and implementation strategies because they depend on the specific industry and project, so this must be done directly by the project implementation team. For sure, balancing solution related to the mitigation hierarchy, such as if it is more financially sustainable an offset or restoration solution instead of an avoidance or minimization also depends on each project. In the following section we gathered strong principles and instructions for implementation.

Project Sustainability Assessment
When evaluating the feasibility of a new project, the first step is also to assess if sustainability goals are met, and at which grade. The Key Performance Indicators (KPIs) used for the initial assessment will also lead the sustainability monitoring and controlling throughout the project life cycle. The assessment must be sized according to the dimension and the level of risk of the project, keeping in mind that the effort will be not only required in the preliminary assessment, but also all along the project life cycle, for managing, monitoring and controlling the indicators assessed in the preliminary phase.

International Financial Institutions provide guidelines for assessing sustainability in projects financed by them [19–23]. Some authors suggest maturity models for project teams and performing organizations [7]. Sometimes it is recommended to draw a sustainability log of the project (similar to a risk log including sustainability goals and related actions), and, for each goal, provide a qualitative rating and, in case, preventive and corrective actions [3]. Moreover, according to the Environmental and Social Policy of the European Bank for Reconstruction and Development, almost all infrastructure megaprojects "could result in potentially significant adverse future environmental and/or social impacts which cannot readily be identified or assessed" [19] and need specific extra effort for assessing and overcoming the negative impact.

Sustainability Rating Systems
In literature, several authors suggest to apply a rating system to establish how much a project is addressing sustainability goals. Rating systems basically consist in a qualitative or semi-quantitative analysis of a project, evaluating several criteria and providing points for each. These are a kind of "rule of thumb", very effective practically, because they relieve people in field to think about high level values for sustainability, but provide a very practical and immediately applicable guidelines to determine sustainability level on the project.

What is currently missing is the development of sustainability rating system suitable for evaluating any kind of project, independently from the industry it belongs.

In fact, in the last decade, several rating systems have been set up and applied to address sustainability in the civil and construction industry. These rating systems consist basically in a qualitative analysis of projects, evaluating several criteria and providing points ("credits") for each. The level of sustainability is generally expressed according to thresholds like "Gold", "Silver" etc., according to the number of credits provided to the project. The maximum number of credits potentially given vary from one criterion to another, so that each criterion has a different weight and concur differently to the total. Currently there is no rating system able to address indifferently building and infrastructure projects, reflecting the specialization that the two industries have. Moreover, there are several differences from a rating system to another when addressing the same project [16, 17]. The most used in 2018 are LEED by Green Building Council for building projects and Envision by Institute for Sustainable Infrastructure for infrastructure projects. Both require that professionals willing to apply the rating system get certified according to a proprietary certification system, and maintain the certification over the years [33, 34].

The application of rating systems shortcut the problem of having clear sustainability criteria expressed in the organization strategy, and a clear budget for sustainability, because rating criteria automatically become project constrains, that are easier and easier to address by project management people than high level declaration in the business case or project charter, with no or little related funds. Rating systems also help in implementing sustainability when personnel is not strongly trained on it.

Sustainability-Related Documentation, Tools and Techniques in Project Phase: Sustainability Management Plan, Sustainability Breakdown Structure, Sustainability Log

In this section we will go through the way sustainability is supposed to be managed in project phase, as this is relevant to estimate the effort required.

If project size allows it (and this is the case for any Megaproject), a Sustainability Management Plan can be drafted. The aim of such plan is "Requiring the development of an environmental and social management system, as a dynamic, adaptive, and continuous process, initiated and supported by the promoter's senior management, while fostering meaningful communication and dialogue among the promoter, its workforce, local communities and, where appropriate, other stakeholders. The system should be commensurate to the size and nature of the project activity" and "Assigning actions and responsibilities, including resources, key performance indicators, funds, skills, etc. to implement the measures" [20]. The same position is lined out by the European Reconstruction and Development Bank: the Suitability Management Plan, called Environmental and Social Management Plan (ESMP) "will define desired outcomes as measurable events to the extent possible with elements such as targets and performance indicators that can be tracked over defined periods" [19]. To be effective, it must be very operation-oriented and be deeply integrated in the other project plans. In a PMBOK environment, it should be included or being and auxiliary plan of the Project Management Plan.

According to [19] "The ESMP will reflect the mitigation hierarchy and, where technically and financially feasible, favour the avoidance and prevention of impacts over minimisation, mitigation or compensation". Similarly [20], such document should be focused on:

Table 2.1 The mitigation hierarchy according to [35]

Avoidance	Measures taken to anticipate and prevent adverse impacts on biodiversity before actions or decisions are taken that could lead to such impacts
Minimization	Measures taken to reduce the duration, intensity, significance and/or extent of impacts (including direct, indirect and cumulative impacts, as appropriate) that cannot be completely avoided, as far as practically feasible
Restoration	Measures taken to repair degradation or damage (…) following project impacts that cannot be completely avoided and/or minimized
Offset	Measurable conservation outcomes, resulting from actions applied to areas not impacted by the project, that compensate for significant, adverse impacts. of a project that cannot be avoided, minimized and/or restored

- "prevent the negative impacts that could be avoided;
- mitigate the negative impacts that could not be avoided but could be reduced;
- compensate/remedy the negative impacts that could neither be avoided nor reduced; and,
- enhance positive impacts."

The mitigation hierarchy has several expression across literature, most of times represented in a pyramid where the most desiderable (avoidance) is on the top, and the least desiderable (offset) on the bottom. We will refer to [35] (Table 2.1).

It is of basic importance to point out that cost faced by the project for the implementation of mitigation strategies are mostly due to the internalization of externalities, so that, in evaluating financial sustainability of a project, these costs and benefits must be carefully weighted. It is also important that, in the evaluating phase, the promoter of the project looks for benefits related to the implementation of sustainability practice to its project, and look for beneficiary entities for sharing costs.

Lastly, for Sustainability deployment, in order to assess the business value of implementing sustainability, the design process must be able to define, collect, track and analyse relevant data. Sustainability can be managed at the same way Risk Management is operationally managed, drafting a Sustainability Breakdown Structure, and issuing a Sustainability Log including specific action to be executed to get each sustainability goal. It has to be managed as, aside the classical "iron triangle", also referred as "triple constrain", there would be another "sustainability iron triangle" having the 3P at its vertexes and encompassing the classical "iron triangle" into it, as a lower level of the big picture. It must be built focusing on the combined goals of 3P sustainability.

A criticality could arise if sustainability is managed separately from other project activities, becoming a bureaucratic level, little integrated and separately managed from other project activities, so that it is important that it is included into integration management and project management plan (Table 2.2).

Since the initiating phase of the project, a sustainability skills assessment must be performed to people involved in the project at all levels. The awareness of what sustainability 3P implies must be spread at all levels to ensure the sustainability implementation to be effective. Training sessions also represent costs on the project

Table 2.2 An example of sustainability breakdown structure to be customized according to the specific project. In some case, a 5P framework could be used

	Planet	• Water • Energy • Pollution • Mobility, transportation and logistics • Wildlife • Materials reduction and reuse • (...)
Sustainabilitybreakdown structure	People	• Ethics, equity and rights • Labour, health and safety • Transparency • Wide stakeholder identification and engagement • Development of skills and capabilities sustainability related • Development of skills and capabilities management related • Sustainable procurement and contracting policy (partnership) [36] • (...)
	Profit	• Corporate strategy alignment [37] • Return of investment • 3P LCA • Externalities internalization • Flexibility and adaptive approach • Long term adaptability • Risk management • (...)
	(Peace)	
	(Partnership)	

that must be assessed and included into the business case. Management skills should also be addressed: even if not directly related to sustainability, they are critical for implementation as sustainability requires the integration of several aspects coming from different disciplines and sectors, so that having personnel skilled in management is a keystone for sustainability implementation.

For the sustainability approach to be effective, all the supply chain must be sustainable [38]. This will also cause extra effort for procurement people to identify, assess, and, in case, train suppliers.

In project closing, retrospectives and lessons learned sessions must be organized, that will benefit the performing organization future projects.

2.4   Findings and Next Steps

According to the most recent literature review, it results critical that sustainability is accurately evaluated in a 3P framework. Integration of the 3 pillars in a "sustainability iron triangle" or "sustainability triple constrains" is the basis to achieve concrete sustainability. Without integration, that is keeping separated social and environmental plans from the economic or financial plan, sustainability will stay a fair topic to talk about without any concrete opportunity to become real.

In the second section, we listed detailed areas where to look for costs and benefits, in order to help business analysts and, in general, personnel defining the feasibility of projects, not to forget any aspect of a project, in order to draft robust business cases. In particular, sustainability-related activities, documents and training have been identified.

At the end of the analysis, it becomes clear that:

1. Sustainability, to be achieved, requires relevant effort and resource expenditure, that must be encompassed in the feasibility evaluation, and compared with benefits that it provides;
2. To be properly evaluated, benefits of sustainability must include the internalization of externalities that in traditional business cases are not included: if not approached in this way, sustainability will go on appearing as a pure cost in project budget and, from the project manager perspective, a cost to be reduced and cut any time the opportunity arise;
3. Strategic goals coming from the performing organization or from the funding institution in case of SPE must include clear and detailed sustainability goals, in order to let the business analyst include a detailed budget for sustainability-related activities in the feasibility study;
4. Rating systems, when applicable, are good "rule of thumb" for facilitating sustainability goals to be achieved at project level, transforming sustainability goals into project constrains having specific budget;
5. All IFIs have guidelines for assessing sustainability of their funded project, often leaving environmental and social aspects untied to economic and financial goals;
6. Management in general and project management in particular is a critical competence for implementing sustainability strategic goals of organizations: specific tools and techniques must be included in the project management frameworks to integrate the three pillars of sustainability into a cohesive whole, a sustainability triple constrains.

The research is at the starting point. We based this work on literature review, direct experience in managing projects and debate within MeRIT research group. In the following stages we will look for information from practitioners and organizations representing stakeholders of a megaproject pipeline (Policy makers, Funding entities, Consulting companies, NGI/O, EPC, SPEs, Operations entities, Regulators, Local communities), developing case studies, performing interviews, questionnaires, focus groups and round tables, looking for best practices in integrating and achieving 3P sustainability in megaprojects.

References

1. Elkington J (1997) Cannibals with forks: the triple bottom line of 21st century business. Capstone Publishing, Oxford
2. Tharp J (2012) Project management and global sustainability. In: Proceedings of PMI® Global Congress 2012—EMEA
3. VVAA (2016) The GPM P5 standard for sustainability in project management release 1.5. Green Project Management
4. Aarseth W et al (2017) Project sustainability strategies: a systematic literature review. Int J Proj Manag 35:1071–1083
5. VVAA (2015) Transforming our world: the 2030 agenda for sustainable development. Resolution of the General Assembly of the United Nations
6. Brook JW, Pagnanelli F (2014) Integrating sustainability into innovation project portfolio management—a strategic perspective. J Eng Technol Manag 34:46–62
7. Silvius G et al (2017) Sustainability in project management. Routledge, London
8. VVAA (2017) Guide to the project management body of knowledge, 6th edn. Project Management Institute
9. Silvius G, Ron Schipper R (2015) A conceptual model for exploring the relationship between sustainability and project success. Procedia Comput Sci 64:334–342
10. Marcelino-Sadaba S et al (2015) Using project management as a way to sustainability. From a comprehensive review to a framework definition. J Cleaner Prod 99:1–16
11. Martens ML, Carvalho MM (2017) Key factors of sustainability in project management context: a survey exploring the project managers' perspective. Int J Proj Manag 35:1084–1102
12. Carvalho MM, Rabechini R (2017) Can project sustainability management impact project success? An empirical study applying a contingent approach. Int J Proj Manag 35:1120–1132
13. Gimenez C, Tachizawa E (2017) Extending sustainability to suppliers: a systematic literature review. Supply Chain Manag Int J 17(5):531–543
14. Martens ML, Carvalho MM (2016) The challenge of introducing sustainability into project management function: multiple-case studies. J Clean Prod 117:29–40
15. Sánchez MA (2015) Integrating sustainability issues into project management. J Clean Prod 96:319–330
16. Chandratilake SR (2013) Sustainability rating systems for buildings: comparisons and correlations. Energy 59(15):22–28
17. Oluwalaiye O, Ozbek ME (2019) Consistency between infrastructure rating systems in measuring sustainability. Infrastructures 4:9
18. Økland A (2015) Gap analysis for incorporating sustainability in project management. Procedia Comput Sci 64:103–109
19. EBRD (2014) Environmental and social policy. European Bank for Reconstruction and Development
20. EIB (2018) Environmental and social practices handbook. European Investment Bank
21. IDB (2019) Modernization of the environmental and social policies of the IDB. Inter-American Development Bank
22. WB (2017) The World Bank environmental and social framework. The World Bank
23. ADB (2019) Operational priority 3: tackling climate change, building climate and disaster resilience, and enhancing environmental sustainability, 2019–2024. The Asian Development Bank
24. Epstein MJ, Rejc Buhovac A (2017) Making sustainability work, 2nd edn. Routledge, London
25. Banihashemi S et al (2017) Critical success factors (CSFs) for integration of sustainability into construction project management practices in developing countries. Int J Proj Manag 35(6):1103–1119
26. Beske P et al (2014) Sustainable supply chain management practices and dynamic capabilities in the food industry: a critical analysis of the literature. Int J Prod Econ 152:131–143
27. Chong H-Y et al (2017) A mixed review of the adoption of Building Information Modelling (BIM) for sustainability. J Clean Prod 142:4114–4126

28. Clinning G, Marnewick C (2017) Incorporating sustainability into IT project management in South Africa. S Afr Comput J 29(1):1–26
29. Shah S, Naghi Ganji E (2018) Sustainability adoption in project management practices within a social enterprise case. Manag Environ Qual Int J
30. Terrapon-Pfaff J et al (2014) A cross-sectional review: impacts and sustainability of small-scale renewable energy projects in developing countries. Renew Sustain Energy Rev 40:1–10
31. Xue B, Liu B, Sun T (2018) What matters in achieving infrastructure sustainability through project management practices: a preliminary study of critical factors. Sustainability 10:4421
32. UNEP (2009) Design for sustainability—a step-by-step approach. United Nations Environment Programme
33. Institute for Sustainable Infrastructure home page. www.sustainableinfrastructure.org. Last accessed 2019/12/1
34. World Green Building Council home page. www.worldgbc.org. Last accessed 2019/12/1
35. Ekstrom J et al (2015) A cross-sector guide for implementing the mitigation hierarchy. Cross Sector Biodiversity Initiative
36. Peenstra R, Silvius G (2017) Enablers for considering sustainability in projects; the perspective of the supplier. Procedia Comput Sci 121:55–62
37. Silvius G et al (2017) Considering sustainability in project management decision making. An investigation using Q-methodology. Int J Proj Manag 35:1133–1150
38. VVAA (2010) Supply chain sustainability—a practical guide for continuous improvement. UN Global Compact Office and Business for Social Responsibility

Chapter 3
Management for Stakeholders Approach for a Socially Sustainable Governance of Megaprojects

Barbara Barabaschi

Abstract The purpose of this paper is to provide a brief overview from a multidisciplinary literature (organizational, political, sociological) on how to manage megaprojects in contemporary societies, focusing on the role of stakeholders engagement. Starting from the main points of weakness stated in literature on megaprojects failure, the paper analyzes these points from a sociological and political perspective in order to understand how to overcome these limits and increase the probability of megaprojects to succeed. Among these points the multi-stakeholder nature of megaprojects, the kind of actors involved and the difficulty to engage all of them to assure an inclusive participation along with a socially responsible model of governance, are emphasized. This conceptual paper refers to both social network and stakeholder theories to integrate the current theoretical body of literature in the field of project management. In particular, the approach of management-for-stakeholders is presented to complement the more traditional management-of-stakeholders. It allows to emphasize megaproject social responsibility and to promote a more comprehensive understanding of megaproject governance in a context of sustainable development.

Keywords Stakeholder management · Megaproject governance · Stakeholder engagement · Theoretical analysis

3.1 Introduction

Megaprojects have been broadly described in economic terms as "large-scale, complex investments that typically cost a billion dollars and up, take many years to develop and build, involve multiple public and private stakeholders, are transformational and impact millions of people" [1]. This paper highlights the complexity that marks out a megaproject, in particular, the impact of social dimension through the

B. Barabaschi (✉)
Università Cattolica del Sacro Cuore, via Emilia Parmense, 84, 29122 Piacenza, PC, Italy
e-mail: barbara.barabaschi@unicatt.it

© The Author(s), under exclusive license to Springer Nature Switzerland AG 2020   27
E. Favari and F. Cantoni (eds.), *Megaproject Management*,
PoliMI SpringerBriefs,
https://doi.org/10.1007/978-3-030-39354-0_3

investigation of the stakeholders role and the relevance of a sustainable governance of them. The intricacies arise from the politics associated with funding, managing and governing complex social and organizational relations.

Some scholars, among economists in particular, (see for example, [2]) suppose a separation between economics and ethics spheres, while others (such as [3]) affirm that every organization theory incorporates a moral dimension, even if it is most of the time implicit. The sociological perspective aims to underline that actors are embedded in a relational system and it is necessary to take into consideration these relations to understand their behavior [4]. In particular, relations between firm and stakeholders are based on moral commitments, not only to optimize profit, but also to manage stakeholders relationships in an optimal way. As a consequence, these relations can be valuable for the company as a reflection of its values and principles. Each firm should define fundamental moral principles and use these principles as a basis for decision making.

Evan and Freeman [5] in defining their normative theory of stakeholder management suggest two principles that managers should follow to coordinate stakeholder interests also when managing a megaproject:

- principle of corporate legitimacy. The company should be managed for the benefit of its stakeholders. Stakeholders must participate in decisions that substantially affect their wellbeing;
- stakeholder fiduciary principle. Managers must act in the interests of all stakeholders as their aim coincide with those of the firm to ensure its survival.

The stakeholders' theory remains ambiguous concerning its foundations and presents certain number of limits. On the one hand, it joins in a relational representation of the organization based on contracts, which suppose that the conflicts of interests can be solved by insuring a maximization of each group interests [6], on the other hand stakeholder relations have also a social nature and not always actors behavior is guided by rationality. At the same time, literature on megaprojects shows a growing interest for more ethical and sustainable projects and a conscious endeavor for fairness and engagement of all stakeholders through a management-for-stakeholders' approach [7].

Other trends that complete the challenging scenario in which megaprojects are designed and implemented, concern a greater sensitivity towards the environment, the increased social investing and the information technology revolution that has significantly changed the way to work, emphasizing the role of knowledge in every context. These trends have added complexity to stakeholder relationships, but at the same time, they have allowed project to be more inclusive and social value oriented. Management-for-stakeholders tries to summarize all the trends cited, that is why it seems worth to better investigate this approach.

3.2 Management-*of*-Stakeholders Versus Management-*for*-Stakeholders

Stakeholder theories trace the difference between the more traditional approach of management-of-stakeholders and the newer management-for-stakeholders [8]. These two approaches assign a different meaning to the term "stakeholder" and are based on different values and principles [9].

Management-of-stakeholders considers stakeholders as providers of resources [10] and project managers main aim is to make the stakeholders comply with project needs [9]. As a consequence, actors having more resources to offer to the project will receive more attention with respect to those having a lower level of resources. This approach has received criticism for being too opportunistic and not ethical. Indeed, a number of projects have failed not only for economic reasons, but also due to the protests of groups or associations not engaged and not considered as relevant stakeholders by project managers. Involvement ranges from committed stakeholders like investors and civic authorities to those that are reluctant, embedded in existing communities, such as social movements and advocacy associations. That is why practitioners and then scholars started to underline the need for managers to consider stakeholders engagement as a priority. So a megaproject management requires to govern and shape relationships among the various actors involved, directly and indirectly (financiers, employees, customers, communities, etc.), and not only to arrange a series of exchanges of material resources to maximize the shareholders profit. Scholars noted that it is convenient for managers to pay attention to actors that also indirectly could influence project performances, such as activists, regulators and other representatives of the community [7]. This is the essence of the so-called management-for-stakeholders, a more holistic approach than the classic and instrumental management-of-stakeholders (e.g. [11]).

Management-of-stakeholders approach requires asking the question: "How should we distribute the burdens and benefits of corporate activities among stakeholders?" Besides, management-for-stakeholders asks the question: "How can we create as much value as possible for all the stakeholders?" In the first case the priority is the allocation of the value produced by the project among primary stakeholders, while in the second case it is the creation and sharing of the value with all the actors involved. It is based on the understanding that all stakeholders are valuable in their own rights regardless of their direct and concrete contribution to the project [12]. In particular, no stakeholder stands alone in the process of value creation, stakeholder interests are tightly tied.

Moreover, the contemporary global purpose to act a sustainable development calls for paying attention to specific values typical of management-for-stakeholders, such as transparency, fairness, trust. In this sense, this approach allows to combine the specific project goals with others expression of the common good, so more general goals, but still crucial to assure project success. Freeman et al. [7] noted that this approach offers an inclusive perspective which aims to involve a broader group of stakeholders, by meeting or exceeding their needs and expectations in coherence with a project socially responsible governance which finally attempts to balance its economic, ecologic and social interests [13].

Managing for stakeholders requires that the interest of all groups of stakeholders have to go together over time. This is clearly more difficult than choosing a group as a priority. As already noted, the managing for stakeholder mindset asks how it is possible to create and share the project value with all the stakeholders, since this is the best way to produce long-term results [7]. Managerial practice, in particular, has demonstrated that in the long-run it is no possible to obtain benefits for a stakeholder without caring for those of others.

Furthermore, when there is dissonance, it is necessary to find a reframing of the basic business proposition so that more stakeholders win continuously over time. Stakeholders that are difficult to please or are in conflicts on the project guidelines can be in any case sources of value creation, if approached with the "no trade-offs" mindset of managing-for-stakeholders. Rather than give into trade-offs, practitioners suggest first try to reframe the basic questions such as "How can we improve the project value for all the stakeholders?" or "Have we engaged all the actors having an interest in the project?".

However, scholars stressed some potential drawbacks in management-for-stakeholders approach, such as the risk of losing focus on those stakeholders vital for the project's survival, or it may lead to escalating stakeholders expectations without being able to guarantee to satisfy them [14]. In addition, even if management-for-stakeholder approach recognizes as important the rights of each actor, due to limited resources, project managers cannot always address the concerns of every potential stakeholder, in practice they tend to define a stakeholder classification and a priority schema to follow in satisfying their demands. That is why, within the wider debate on the narrative of contemporary capitalism, literature moves from the idea to combine management-for-stakeholders and management-of-stakeholder considering these approaches as complementary [15]. The classic narratives of capitalism, to tell the story of value creation has assumed the perspective of just one stakeholder, in turn, worker, government, investor, manager, entrepreneur, but it is clear that this approach is inadequate to represent the wider spectrum of actors interested in a firm and who have an impact on its survival in the long run. Final aim should be to develop a superior stakeholder theory within the wider debate on the narrative of contemporary capitalism [16]. Nevertheless, actually, this ambitious aim is not yet realized, so the more persuasive theoretical reference remains the stakeholder-for-management approach. The stage of the debate is summarized by Freeman et al. [7] "we should re-frame capitalism in the terms of stakeholder theory so that we come to see business as creating value for all the stakeholders".

Freeman et al. [7, 8] identifies normative strategies at enterprise level and envisions a variety of options that managers can choose in order to improve firm unique situation to the environment needs and features. Though there is not a specific answer as to how management should choose among these options, Freeman suggests that there should be a "fit" between stakeholders, values, social issues and the society within which managers operate. A way is to organize extensive planning and strategic activities, including stakeholder audits, to identify various social issues and values, as well as to help monitor performance and keep score with stakeholders over time.

Although the literature is moving forward, there has not been an academic effort to identify and summarize the underlying assumptions that make the management-for-stakeholders approach beneficial (or not) to megaproject performance and scarces are studies explaining how to traduce this approach into practice. This paper wants to contribute to fill this gap with reference to some aspects often overlooked in literature, because conceived as complementary in the traditional management-of-stakeholder approach. These aspects refer to: secondary actors involvement; community of practices approach for stakeholders communication and social responsibility monitoring throughout the megaprojects lifecycle.

3.3 The Involvement of Secondary Actors

Literature (see [17, 18]) distinguishes stakeholders into two categories: primary and secondary stakeholders. The first refers to those stakeholders which engage in formal contractual relationships with a company, such as customers, employees and shareholders. This kind of actors has the greatest and more direct impact on the project, often their collaboration is essential to assure the project success (i.e. investors, workers, …), consequently managers devote great attention to their needs. However, stakeholder theory also stresses the importance of *secondary stakeholders* [17], which do not engage in transactions directly relating to the company's going concern and lack formal contractual relationships (examples are, public administration, mass media, environmentalist associations, religious groups and other non-governmental organizations). Clarkson underlines the interdependence between the two groups of stakeholders. Indeed, there is growing evidence of the power of secondary stakeholders able to induce companies to respond to their needs. Also recent studies (see [13, 19]) highlight the beneficial impact on project performance deriving from managing secondary, but legitimate, stakeholders. In particular, actions addressed to this group will help managers to reduce planning misjudgment and to increase transparency and accountability in the project decision making process. For example, due to the unavoidable impact of a megaproject on both people and places, it is suggested that seeking local community opinions in the initiation phase of the project and monitoring the megaproject impact at the local level can help to improve final performances [20, 21].

The main question, at this point, concerns how to assure a useful engagement of both primary and secondary stakeholders.

Project management has focused strongly on those actors able to control project resources, whilst the effect on the legitimate 'secondary stakeholders', such as the local community, remains widely underestimated. According to Reed [22], stakeholders typically only get involved in decision making at the implementation phase of the project cycle, and not in earlier project identification and preparation phases.

Managerial implications have to consider the impact of engaging with a broader number of stakeholders from the initiation phase of the project, so to envisage an organizational schema in which each stakeholder group is represented and has the possibility to express his opinion. Evaluating the practical implications of including secondary stakeholders' inputs in the decision-making process during the initiation phase of megaproject, is another relevant step allowing to enhance project performance.

In order to find the optimal strategy for all groups of stakeholders, scholars have proposed different criteria to map them, each classification to be adapted to the specific context in which megaproject is developed. The aim of these taxonomies is to go beyond the dichotomy between primary and secondary stakeholders in order to pass from theory to practice. Just as an example, three proposals will be cited. One has been elaborated by Freeman (the most convinced scholar of the management-for-stakeholder approach) who suggests analyzing the stakeholder behavior and possible coalitions among groups thorough investigating in the past actions of such kind of groups. It is necessary to analyze the actual behavior of stakeholders, their cooperative potential and competitive threats. Savage et al. [23] gave guidance on how to measure these variables. The power of threat is determined by the type and the entity of resources that stakeholders have, their ability to build coalitions and relevance of the threat for the project. The potential to cooperate depends on how close the stakeholder interest is to the project organization: the closer is the link, the higher is the interest to cooperate. As a result of these two variables combination, it is possible to identify four types of stakeholders and the relative strategy that managers could adopt to optimize the relations with each group. Supportive stakeholders require an offensive strategy, that implies trying to reinforce stakeholders view or perception of the project to maximize the convergence to the organization aims and values. Non supportive stakeholders are addressed a defensive strategy in order to prevent potential threats by reinforcing beliefs and values of the project organization and finally favoring stakeholders integration. When managers deal with mixed-blessing stakeholders a swing strategy is useful in order to adapt rules or transaction process to the stakeholders current needs and features. Marginal stakeholders require a hold strategy and so managers continue their program.

Friedman and Miles [24] propose a similar schema to classify stakeholders and deduce guidelines for managers action. In this case the variables considered are necessary or contingent relationships between managers and stakeholders. The first are internal to a social structure or to a set of logically connected ideas. Contingent relations are not integrally connected. In this case, five strategies are suggested. A defensive strategy to reinforce relationships for necessary contingent relations among actors all having something to win; a defensive strategy to discredit oppositional views, or to eliminate stakeholders having opposite interests and not available to reach a compromise. An opportunistic strategy is suggested when relations are necessary contingent, but actors involved are not directly connected (for example, a company and a public authority). Finally, a compromise strategy is the best way to follow in the case of necessary incompatible relations, that is when all actors interests are related to each other, but their actions could damage the relationship itself.

The last study cited is that of Mitchell et al. [11]. It has become the major contribution on strategies to govern relationships between managers and stakeholders. They put forward three criterions to define a hierarchy of stakeholders: the stakeholders power to influence the project, the legitimacy of the stakeholders relationship with the project and the urgency of the stakeholders claim with managers. The combination of these three criterions leads to seven stakeholder types: dormant; discretionary; demanding; dominant; dangerous; dependent; definitive. A stakeholder is considered a low priority if it has only one attribute; he becomes a moderate priority when it is linked to two attributes and a high priority if three attributes are recognized. The possession of an attribute is subjective, because it depends to the manager opinion. Furthermore, this possession is also dynamic, that is a stakeholder may be considered legitimate in the first phase of a project but become urgent at a later time with the media support and then powerful with a boycott campaign.

The decision making of megaprojects is typically not driven by the real needs of society, but only by the technological, political, economic and aesthetic sublimes presented by Flyvbjerg [1, p. 8] which "ensure coalition between those who benefit from these projects and who will therefore work for more such projects".

Unpopularity and local opposition are common threat for megaprojects whereby secondary and external groups try to influence the projects implementation [25].

Söderlund [26] and Bakker [27] assert that temporary organization approaches see projects as social systems, whereby behavior (not just decision-making) through social interactions is highly influenced by the context in which they are embedded. Projects are temporary and unique, so additional efforts are needed to disseminate trust among the project stakeholders [28]. Consequently, a deep understanding of the cultural, organizational and social environments surrounding projects is crucial. Therefore, managers of major programs have to analyze the concerns, needs and moral issues of (local) stakeholders, not only at the inception phase of the project, but throughout its entire lifecycle.

Interesting theoretical tools to implement management-for-stakeholders come from the actor-network theory and from the sociology of translation (see [29]) which help to understand how networks emerge and are transformed through processes of translation and power relationships.

Following the actor network theory (see, for example, Selznick [30] a project's success largely depends on how the stakeholders are engaged and their expectations from the project are being fulfilled. Hence, it is important to evaluate the complexity of such megaprojects and to explore the underlying relational networks of multiple stakeholders that influence the delivery of the expected outcomes. Translation process consists of different stages. The first is "problematization", actors identify and involve a number of actors whose roles and relationships configure an initial problem-solving network [31]. The second is "interessement", actor involves others sufficiently to agree with its proposal [29]. Through this process, those supporting the emerging network incite actors into fixed places and weaken the influence of other actors that may disestablish the developing network [31]. Interessement does not necessarily lead to successful alliances and eventually translations; it needs to be reinforced by enrolment, that is the third stage and consists of negotiations among the actors target for enrolment, but also with those actors who can potentially threaten network stability [29].

The networks are continuously evolving and transforming through processes of translation in which a temporary actor-network progressively takes form. Those playing the role of the controlling actors (in the case we are studying, project managers and representatives of each stakeholder groups) develop different strategies to drive the translation in order to enroll and mobilize other actors [32]. During a successful translation, those being controlled (stakeholder groups) have to remain faithful to the objectives of those who control and those exerting control are given the right to represent those mobilized. Nevertheless, translation processes are not always successful. When managers or stakeholders representatives fail to get other actors to comply with them, a process of dissidence takes place [29] and it is necessary to activate a new translation process through negotiation and adjustment mechanisms to find a new equilibrium among the various stakeholders groups or between stakeholders groups and managers. The process of representation takes place using chains of intermediaries who "little by little reduce the number of representative interlocutors" [29], that are influenced by power relationships established within the actor-network. The sociological analysis allows to understand how a few obtain the right to represent the many silent actors within the social entities (stakeholders groups) they have mobilized and described the way in which actors are associated and oriented to remain faithful to their alliances. These aspects are both crucial to assure the translation process and a fruitful stakeholders network governance. This latter is also influenced by an effective communication among the various actors involved. To this factor is devoted the next paragraph.

3.4 Community of Practice Approach for Stakeholders Communication Strategies

Communication issues permit to move the analysis to an even more practical level. Indeed, it offers the lens to a theoretical approach able to integrate that of the actor network theory to compensate some limits that it presents, such as it could take for granted an informative symmetry among different actors, or it could ignore the specificities of the context in which relations occurred, or could apply the translation concept regardless of the empirical data collected. The practice-based approach (PBA) is presented in literature as an option to integrate the actor network theory results. It began in the thinking of Bourdieu and Giddens (as well as Heidegger, Wittgenstein, and Garfinkel) its success is reflected in its application in fields such as strategy (e.g. [33]); knowledge sharing, learning and communities of practice (see for example, [34–36]). One of the core themes of PBA is the notion of communities of practices, a construct useful to explain and traduce into practice the management-for-stakeholder approach in megaprojects.

Communities of practice that regularly bring together people who share areas of interest have been theorized by Wenger [36] and differs significantly from that of work teams more diffused in management-of-stakeholders approach. While teams are formed by managers and respond to specific deliverables, communities of practice emerge from a voluntary basis and can be seen as a way to enhance individual and collective competencies through the sharing of a common repertoire of resources. It is from this shared practice that a community's member relies on the knowledge capitalized by the community to carry out further activity. The principles of inclusiveness, trust and social cohesion at the basis of communities of practices are in line with the ones sustained in management-for-stakeholders approach to govern megaprojects. These latter can be seen as social entities, a set of relationships among multiple interrelated stakeholders networks constantly crossed by information flows, essential to maintain vital the networks themselves and to create value for the project. This value is not merely instrumental for work, but it helps to share a body of common knowledge, practices and approaches, it increases the sense of belonging, finally developing a common identity [37]. As a result, communication should be based on trust-facilitating behaviors to promote stakeholders confidence during all phases of the project lifecycle. Communities of practices allow the various stakeholder groups to assume the organization of a critical asset, to understand what knowledge will give them a competitive advantage in the different phases of the project. They then need to keep this knowledge up to date, deploy it and spread it across the entire organization. There may be disagreement but it is through a process of communal involvement, including all the controversies, that the knowledge useful to assure the project equilibrium and its development is developed. The involvement of different views, in this case is considered as an enriching factor that complement managers knowledge. This collective characteristic of knowledge does not mean that individuals don't count. In fact, the most inclusive and cohesive communities welcome strong personalities and encourage disagreements and debates. It can be useful to create a charter that covers why the community exists, who the members are, the community's goals and principles for how it will run.

A specific role is assigned to the information system which takes advantage from the use of new technologies in order to promote transparency and accountability of the megaproject activities and outcome through widespread communications between stakeholders and managers.

Literature counts series of studies using actor network theory with a practice-based approach to better understand the adoption of technologies in information processes. Such studies have focused on technologies such as intranets, electronic work time registrations systems; geographical information systems; health management information systems [38]. Others have investigated the use of new media on social network theory beyond classic social methods describing a series of opportunities. Hyperlink networks, for example, in which the nodes are websites and the ties are the hyperlinks that connect them, may be analyzed to trace the diffusion of content between mainstream media and blogs, or to determine the extent to which prominent mainstream media versus bloggers wield influence in media and public agenda-setting [39]. The development of models that allow to consider multiple types of ties, actors and multilevel networks enables consideration of greater complexity in the study of diffusion and mediated social influence, particularly in the context of megaproject. These developments are relevant in the contemporary knowledge society in which actors may be both producers and consumers of information and may access content from many different types of sources and using many different types of media.

All studies cited have demonstrated their efficiency in promoting online communities and the convergence of members towards a common goal. To assure that this happens and that the stakeholders accept the new technologies as mediator of their relations, literature reveal the crucial role of powerful actors whose strategies can mobilize other actors. The adoption of technologies can be highly affected by the lack of negotiations between relevant actors [40], functional to continuous translation processes.

In line with the management-for-stakeholders approach, from some studies has emerged that actors situated outside the boundaries of a particular actor-network, and other actor-networks,[1] can be seen as powerful actors that can enforce or weaken a particular network [41]. The introduction per se of an online community does not necessarily lead to its success, but scholars pointed out some requested conditions, in particular, communities need to align themselves to the context surrounding them and to the values of the project's shareholders; aspects such as management sponsorship, extensive promotion and consultation with potential participants are critical for their institutionalization [42]. Indeed another requirement is the ability to use new technology tools for communications. In this sense, the so called digital gap acts as a limit for the participation of all stakeholders. That is why ad hoc training sessions are suggested for all stakeholders (especially the closest to the managers) as well as blended form of communications combining hi-tech instruments and social media, with more traditional tools like journal, bulletin, vis-à-vis questionnaire or interview, etc. Nevertheless, on-line tools are considered a proactive way to really increase inclusiveness and stakeholders participation. In this sense, communication policies constitute a pivotal component of megaproject social responsibility.

3.5 Social Responsibility Monitoring Throughout the Megaprojects Lifecycle

Megaprojects bear extensive and profound social responsibilities throughout their lifecycle. The prolonged life and heterogeneous stakeholders have posed great challenges for the governance of the economic, social, and environmental issues involved [43]. As a consequence ethical and environmental issues, such as risk control [44],

---

[1]In the case of megaproject, actor-network coincide with stakeholders groups.

safety management [45], environmental protection [46] have recently received more attention in the megaproject management literature.

Additionally, the social and environmental aspects of megaprojects dynamically evolve with the advancement of the project lifecycle. A wide variety of salient stakeholders of megaprojects exert distinctive influences on the responses to social and environmental concerns as they have diverse and sometimes mixed motives for the decisions they make and the actions they take. Unlike corporate social responsibility, which rests with specific corporations and usually single individuals (CEOs), megaproject social responsibility can never rest with any single individual or organization. The involvement of diverse actors cooperating closely in order to improve project performance means that an integrated, multi-level systems view is needed to analyze the megaproject social responsibility. The governance of megaprojects is distinctive in its hybridity of both business and politics and regimes of megaproject social responsibility have complex contexts. "Business–Government–Society" (BGS) approach, developed by Steiner and Steiner [47], provides the lens through which to understand social responsibility by analyzing interrelationship and interactions among business, government, and society. In particular, businesses create a megaproject thorough competition and cooperation; government is responsible for creating legislation, while society is involved thorough collective action and participation [48]. The most representative features of megaproject governance are: the duality of leaders (usually one representative of the firm and one representative of the public actor), due to the various actors involved; the plurality of dimensions involved, economic, legal, ethical, ecological, political; the dynamism, because the social networks of stakeholders evolve during the project life course characterized by an high level of uncertainty in the first phases that decrease in the last phases.

The BGS model requires socialization mechanisms to favor the governance of megaprojects, in particular processes of social participation especially with reference to local community and the wider public to be involved in the main project decisions and actions. Social learning is another process deriving from participation as an iterative dynamic to improve project performance. Learning processes are also useful to reduce project complexity and the potential for conflict [49]. Social interaction among all stakeholders is a critical process because it consists in promoting an alliance culture especially with secondary stakeholders, so to assure transparency and accountability of project behavior [50]. This allows to prevent collusion and corruption episodes. Along with continuous social interaction, adjustments to strategies and value management take place in order to adapt to changing situations during the project lifecycle. In addition, the socially responsible performance of megaprojects requires mechanisms of information feedback, blackspot make-up and post-evaluation reports. Finally, social integration is essential to build an effective governance of the system including all issues linked to megaproject social responsibility. Prior research on the social and environmental concerns concluded that megaprojects has been quite fruitful but relatively fragmented. BGS model also intends to encourage business managers and policymakers to rethink their roles and responsibilities in megaprojects and effectively cooperate with society in respect of the social issues. In particular, the government plays a crucial role in assuring independence as protector and coordinator of the public good (the megaproject) and so contributing to reinforce stakeholders trust in the leaders of the project.

Another requirement concerns the skills and competencies of the megaproject team of management. In fact, project team need to have *the right mix of abilities*. Without a well-resourced and qualified network of project managers, advisers and controllers, projects will not deliver the best possible return on investment. Investors and owners need to take an active role in putting together the project team. It is not enough for them to have a vague theoretical overview of how the project should work. They need to create a detailed, practical approach to deal with such likely eventualities as managing quality risks, escalating contractor's costs, or replacing a high-tech supplier. Players must assemble a team that has all the requisite skills, including legal and technical expertise, contract management, project reporting, regulatory approval, stakeholder management, government and community relations, communication skills especially concerning the use of new technology and social media.

Finally, this framework can be adopted within the road map for megaproject practice so as to improve the governance performance of social responsibility and create shared and sustainable value for all stakeholders. The model has the advantage of unifying and coordinating the higher number of variables to count on in megaproject implementation. However, as every systemic approach, it acts as a reference, but at the same time, it requires to adapt the framework to the concrete situation, for example to the nature of the organization and its mission and objectives, the type of project, as well as some contextual factors, such as the level of economic competition, the country in which the organization operates, the government political ideology, the local environment, the institutional context as well as the cultural backgrounds [50]. Indeed, depending on the type of organization (i.e., private, public, hybrid; industry type), both internal and external stakeholders can differ significantly and the organization behavior as well.

3.6 Conclusions and Future Research

At the end of this analysis it is possible to conclude that the management-for-stakeholders approach is able to capture di essence of the contemporary model of capitalism understood as a system of social cooperation and value creation. In particular, the attempts to realize projects in a more ethical way is consistent with the principles of corporate legitimacy and stakeholder fiduciary presented by scholars as the basis of potential project success and mentioned in the first part of this work. The main feature of management-for-stakeholders is that it tries to change the way managers (both private and public) think to the project, that is as a network of actors, not only as a set of resources and this kind of understanding is a strategic requirement to allow ethics to get built into the project and not remaining a mere idea. E. R. Freeman is one of the most fervent defenders of this construct, but also of stakeholder society, a vision that Hutton [51] helped to find and within which stakeholders are the drivers of a new strategic organizational discourse, leading to the formulation of a strategy that is no longer merely responsive but also proactive in its efforts to create greater value for all actors involved.

Organizational strategy frequently fails to achieve the desired results, in fact, historically, megaprojects have performed poorly in terms of benefits, public support and social impact [52]. Starting from this evidence, this study has tried to reinforce the idea that a multidisciplinary and multidimensional approach is needed. In particular, technical and economic perspectives try benefits when integrated to the social one, since focusing too much on technical skills or being too focused on budget leads to inadequate stakeholder management and the omission of the social and political issues so decisive in driving projects value creation and social responsibility in a context of sustainable development.

This paper has also attempted to interpret contemporary trends in megaproject governance through the lens of sociological literature and social research methodology. In particular, the effort has consisted in applying the construct of communities of practices and the actor-network theory along with the practice-based approach to give a concrete representation of management-for-stakeholders presented as a useful answer to the problem of the limited stakeholders engagement practices having caused megaproject failure in the past. It seems that the social network perspective offers invaluable insights to identify and analyze project stakeholders through a holistic perspective.

Finally, this paper presents some limits. It is based mostly on a literature review and developed through theoretical analysis, so further empirical research is needed to support our findings and provide a full understanding of the topic of megaproject societal governance through the management-for-stakeholders approach. For instance, it would be valuable the application of this framework via a case study might be conducted in order to investigate how a particular organization manages its stakeholders during the different phases of the project. Operational issues need to include the availability of network data, stakeholders identification and classification to be included within the scope of the project, definition of the relationship types (i.e. contractual, collaborative, conflictual, etc.).

References

1. Flyvbjerg B (2014) What you should know about megaprojects, and why: an overview. Proj Manag J 45(2):6–19
2. Donaldson T, Preston LE (1995) The stakeholder theory of corporation: concepts, evidence and implication. Acad Manag Rev 20(1):65–91
3. Freeman RE (2008) Managing for stakeholders. In: Beauchamp T, Bowie N, Arnold D (eds) Ethical theory and business, 8th edn. Pearson, London
4. Granovetter MS (1995) Economic action and social structure: the problem of embededdness. Am J Sociol 91:481–510
5. Evan RE, Freeman WM (1990) Corporate governance: a stakeholder interpretation. J Behav Econ 19(4):337–359

6. Fontaine C, Haarman A, Schmid S (2006) The stakeholder theory. Stakeholder theory of the MN. Available at https://pdfs.semanticscholar.org/606a/828294dafd62aeda92a77bd7e5d0a39af56f.pdf. Last accessed 2019/11/24
7. Freeman R, Harrison S, Wicks AC (2007) Managing for stakeholders: survival, reputation, and success. Yale University Press, New Haven, London
8. Freeman RE (2010) Managing for stakeholders: trade-offs or value creation. J Bus Ethics 96(Suppl 1):7–9
9. Eskerod P, Huemann M (2013) Sustainable development and project stakeholder management: what standards say. Int J Manag Projects Bus 6(1):36–50
10. Huemann M, Eskerod P, Ringhofer C (2017) The influence of local community stakeholders in megaprojects: rethinking their inclusiveness to improve project performance. Int J Proj Manag 35(8):1537–1556
11. Mitchell R, Agle B, Wood D (1997) Toward a theory of stakeholder identification and salience: defining the principle of who and what really counts. Acad Manag Rev 22(4):853–886. Available at https://links.jstor.org/sici?sici=0363-7425%28199710%2922%3A4%3C853%3ATATOSI%3E2.0.CO%3B2-0. Last accessed 2019/11/24
12. Aarseth W, Ahola T, Aaltonen K, Økland A, Andersen B (2017) Project sustainability strategies: a systematic literature review. Int J Proj Manag 35:1071–1083
13. Di Maddaloni F, Davis E (2017) The influence of local community stakeholders in megaprojects: rethinking their inclusiveness to improve project performance. Int J Proj Manag 35:1537–1556
14. Eskerod P, Huemann M, Ringhofer C (2015) Stakeholder inclusiveness: enriching project management with general stakeholder theory. Proj Manag J 46(6):42–53
15. Eskerod P, Huemann M, Savage G (2015) Project stakeholder management: past and present. Proj Manag J 46(6):6–14
16. Agle BR, Donaldson T, Freeman RE, Jensen MC, Mitchell RK, Wood DJ (2008) Dialogue: toward superior stakeholder theory. Bus Ethics Q 18(2):153–190
17. Clarkson MBE (1995) A stakeholder framework for analyzing and evaluating corporate social performance. Acad Manag J 20(1):92–118
18. Freeman RE (1984) Stakeholder management: a stakeholder approach. Cambridge University Press, Cambridge
19. Turner R, Zolin R (2012) Forecasting success on large projects: developing reliable scales to predict multiple perspectives by multiple stakeholders over multiple time frames. Proj Manag J 45(5):87–99
20. Olander S, Landin A (2005) Evaluation of stakeholder influence in the implementation of construction projects. Int J Proj Manag 23:321–328
21. Teo MM, Loosemore M (2014) The role of core protest group members in sustaining protest against controversial construction and engineering projects. Habitat Int 44:41–49
22. Reed MS (2008) Stakeholder participation for environmental management: a literature review. Biol Conserv 141:2417–2431
23. Savage GT, Nix TW, Whithead CJ, Blair JD (1991) Strategies for assessing and managing organizational stakeholders. Acad Manag Executives 5(2):61–75
24. Friedman AL, Miles S (2006) Stakeholders: theory and practice. Oxford University Press, New York
25. Boholm Å, Löfstedt R, Strandberg U (1998) Tunnelbygget genon Hallandsas: Lokalsamhallets dilemma [Construction of the tunnel through Hallandsas: the dilemma of the local community]. CEFOS (Centre for Public Sector Research), Gothenburg University, Gothenburg
26. Söderlund J (2017) A reflection of the state-of-the-art in megaproject research: the Oxford handbook of megaproject management. Proj Manag J 48(6):132–137
27. Bakker RM (2010) Taking stock of temporary organizational forms: a systematic review and research agenda. Int J Manag Rev 12(4):466–486
28. Grabher G (2002) Cool projects, boring institutions: temporary collaboration in social context. Reg Stud 36:205–214

29. Callon M (1986) Some elements of a sociology of translation; domestication of the scallops and the fishermen of St Brieuc Bay. In: Law J (ed) Power action and belief a new sociology of knowledge? Routledge and Kegan Paul, London, pp 196–229
30. Selznick P (2011) The old institutionalism meets the new institutionalism. Sociol Perspect 54(3):283–306
31. Linde A, Linderoth H, Raisanen C (2003) An actor network theory perspective on IT-projects: a battle of wills. Action in language, organizations and information systems. Linkoping University, pp 236–250
32. Blackburn S (2002) The project manager and the project-network. Int J Proj Manag 20(3):199–204
33. Chia R, Holt R (2006) Strategy as practical coping: a Heideggerian perspective. Organ Stud 27(5):635–655
34. Gherardi S, Strati A (2013) learning and knowing in practice-based studies. Elgar, Cheltenham
35. Østerlund C, Carlile P (2005) Relations in practice: sorting through practice theories on knowledge sharing in complex organizations. Inf Soc 21:91–107
36. Wenger E (1998) Communities of practices: learning, meaning and identities. J Math Teacher Educ 6(2):185–194
37. Wenger EC, McDermott R, Snyder WC (2002) Cultivating communities of practice: a guide to managing knowledge. Harvard Business School Press, Cambridge
38. Cho S, Mathiassen L, Nilsson A (2017) Contextual dynamics during health information systems implementation: an event-based actor-network approach. Eur J Inf Syst 17 (6):614–630
39. Ognyanova K, Monge P (2013) A multitheoretical, multilevel, multidimensional networkmodel of the media system: production, content and audiences. Commun Yearb 37:66–93
40. Elbanna A (2010) Actor network theory in ICT research: a wider lens of enquiry. Int J Actor Netw Theor Technol Innov 1(3):1–14
41. Avgerou C (2019) Contextual Explanation: Alternative Approaches and Persistent Challenges. MIS Quart 43(3):977–1006
42. Tabak E (2008) Inscription of information behaviour to communities of practice on an organisational intranet. Proceedings of the 20th Australasian Computer-Human Interaction Conference, OZCHI 2008: Designing for Habitus and Habitat, Cairns, Australia, December 8–12 Available at https://dblp.uni-trier.de/db/conf/ozchi/ozchi200
43. Henisz WJ, Levitt RE, Scott WR (2012) Toward a unified theory of project governance: economic, sociological and psychological supports for relational contracting. Eng Proj Organ J 1–2(2):37–55
44. Flyvbjerg B, Bruzelius N, Rothengatter W (2003) Megaprojects and risk: an anatomy of ambition. Cambridge University Press, Cambridge
45. Sun Y, Fang D, Wang S, Dai M, Lv X (2008) Safety risk identification and assessment for Beijing olympic venues construction. J Manage Eng 24(1):40–47
46. Xue X, Zhang R, Zhang X, Yang RJ, Li H (2015) Environmental and social challenges for urban subway construction: an empirical study in China. Int J Project Manage 33(3):576–588
47. Steiner JF, Steiner GA (2002) Business, government and society. McGraw-Hill Higher Education, New York
48. van Marrewijk M, Were M (2003) Multiple levels of corporate sustainability. J Bus Ethics 44(2–3):109–117
49. Sanderson I (2012) Evaluation, policy learning and evidence-based policy making. Public Adm 80(1):1–22
50. Zeng SX, Ma H, Lin H, Chen H (2017) The societal governance of megaproject social responsibility. Int J Proj Manag 35(7):1365–1377
51. Hutton W (1999) The stakeholding society: writings on politics and economics. Polity Press, Cambridge
52. Bruzelius N, Flyvbjerg B, Rothengatter W (2002) Improving accountability in mega projects. Transp Policy 9:143–154

Chapter 4
The Economics of Mega-projects

Silvia Platoni and Francesco Timpano

Abstract Mega-projects play a foremost role not only in developed countries, but mainly in developing countries in several respects: political dynamics, social effects, and economic fallout (see Bovensiepen and Meitzner Yoder in Asia Pac J Anthropol 19(5):381–394, 2018, [4]). As correctly stated by van Wee and Tavasszy in Decision-making on mega-projects—cost-benefit analysis, planning and innovation, pp 40–65, 2008, [31], first, they are heavily under debate at the political level because of their economic impacts and important budget implications; in fact mega-projects are "large-scale, complex ventures that typically cost $1 billion or more, take many years to develop and build, involve multiple public and private stakeholders, are transformational, and impact millions of people" (see Flyvbjerg in The Oxford handbook of megaproject management. Oxford University Press, Oxford, pp 1–18, 2017, [15]). Second, there is a lively debate also among scholars and practitioners owing to the huge cost escalation and schedule delay of these projects (see Flyvbjerg et al. in Megaprojects and risk: an anatomy of ambition. Cambridge University Press, Cambridge, 2003a, [8], Odeck in Transp Policy 11(1):43–53, 2004, [22]), but also due to the uncertainty of their wider economic effects, which can be interpreted as related externalities.

Keywords Evaluation · Valuation · *CBA* · *PESTOL* model

4.1 Introduction: "Over Budget, Over Time, Over and Over Again" (See [13], p. 312)

Mega-projects are crucial to the future of states, cities, and individual livelihoods [16] since they are increasingly used as the preferred delivery scheme for goods and services across a range of businesses and sectors, such as infrastructure, water and energy, information technology, industrial processing plants, mining, supply chains,

S. Platoni (✉) · F. Timpano
Dipartimento di Scienze economiche e sociali, Università Cattolica del Sacro Cuore, Via Emilia Parmense 84, 29122 Piacenza, Italy
e-mail: silvia.platoni@unicatt.it

© The Author(s), under exclusive license to Springer Nature Switzerland AG 2020   43
E. Favari and F. Cantoni (eds.), *Megaproject Management*,
PoliMI SpringerBriefs,
https://doi.org/10.1007/978-3-030-39354-0_4

enterprise systems, strategic corporate initiatives and change programs, mergers and acquisitions, government administrative systems, banking, defense, intelligence, air and space exploration, big science, urban regeneration, and different major events [14].

Nevertheless, mega-projects have become a focus of attention for study and analysis by scholars and practitioners not for this reason, but because they create controversy from the outset and fail to deliver what they promised [1]. In fact the key problem is that these projects often go off the rails either with regard to budget or time, or even both [16, 20]: the performance of mega-projects has seen as problematic in terms of overall too-budget and on-time delivery, as well as in terms of the utility of the mega-project once in operation (i.e. the mega-project do not produce the intended societal benefits).

Atkinson [2] and Zidane et al. [31] claim that the project success need to be defined in at least two dimensions:

1. accomplishing the *result goals* (the project delivery at the completion of the project according to plan),
2. accomplishing the effect goals (the effects of the project, once it has been completed).

Furthermore, Zidane et al. [31] specify that the effect goals can further be categorized into two dimensions:

2.a the effects–benefits for the organization that undertakes the project (*effect goals* strictly speaking),
2.b the effects–benefits for society (*society goals*).

Therefore the success of a project can be looked at and evaluated based on these three major levels of goals [25, 31] and concepts that are applied as criteria for evaluation, such as relevance, effectiveness, efficiency, sustainability, and impact, can be compared to these three major levels of goals.

Indeed, in project evaluation, examinations are basically conducted in view of five evaluation criteria well identified, among the others, in *JICA* [19] and by Zidane et al. [31]:

(1) relevance, defined by Samset [25] as an overall assessment of whether a project is in harmony with the needs and priorities of the owners, the intended users, and other attested parties;
(2) effectiveness, defined by Samset [25] and Zidane et al. [31] as the extent to which the objective has been attained, that is the first-order effect of the project for the users, in the market, in terms of production, etc.;
(3) efficiency, defined in *OECD* [22] as a measure of the ratio between the input and the output, by Samset [25] as the degree to which project outputs have been delivered as planned and in accordance with budget (it could have been done cheaper, more quickly and/or with better quality), and by Zidane et al. [31] as a way of doing things right and producing project outputs in terms of the agreed scope, cost, time, and quality (this last factor often perceived as a by-product of the first three factors);

(4) sustainability, conceived both at environmental and financial level, and defined by Samset [25] and in *OECD* [22] as a measure of whether the benefits of an activity are likely to continue after donor funding has been withdrawn;

(5) impact, defined by Samset [25] and in *OECD* [22] as the positive and/or negative changes produced by a project, directly or indirectly, intended or unintended, in the short and the long term.

The perception of each evaluation criterion is different depending on the timing of the evaluation study [19], as seen in detail in the following two paragraphs.

Mišić and Radujković [20] summarize, and ultimately synthesize, the findings obtained by Cleland and King [6], Pinto and Mantel [24], Pinto and Kharbanda [23], Flyvbjerg [13], Tabish and Jha [28], and Ika et al. [18], who extrapolate the success factors but mainly the failure factors of mega-projects by analyzing the Megaprojects *EU* Cost, the *OMEGA* research, and the *NETLIPSE* network.[1]

The main failure factors identified by those authors are related to (1) ex-ante factors:

(a) the lack of a clear mission, that is do not clearly define the goals and the general direction from the outset [24];

(b) to ignore the project environment (including the stakeholders), do not worry to understand project's trade-off, and do not conduct feasibility studies [23];

(c) the optimism bias, which consists in managers making decisions based on delusional optimism rather than on a rational weighting of gains, losses, and probabilities [13];

(2) mid-term factors:

(d) the strategic misrepresentation, which accounts for flawed planning and decision making in terms of external pressures, that is political pressures and agency issues [13], and which may degenerate to the point political expediency and infighting dictate crucial project decisions [23];

(e) to allow bureaucracy and internal corporate mechanism to be more important than project success [23];

(f) the lack of monitoring and feedback, that is do not timely provide comprehensive control information at each stage of the project implementation [24];

and (3) ex-post factors:

(g) to never admit a project is a failure [23];

(h) the lack of post-failure reviews, losing accordingly the opportunity to learn and understand the main reasons of the failure [23].

All previous failure factors depend on the primordial failure factors, that is:

(i) either to choose a not charismatic and skilled project manager,

---

[1] The main projects of the Megaprojects *EU* Cost Action are the Seville Metro Line and the Zagreb on the Sava River in Croatia, those of the *OMEGA* research are the Lignes à Grande Vitesse (*LGV*) Méditerranée and the Athens Metro Base (*AMB*) Project, and those of the *NETLIPSE* network are the Highway A73-South in The Netherlands and the Lötschberg Base Tunnel in Switzerland.

(j) or to not provide managers, even if prepared, the necessary resources and
     authority/power for project success [23].

Given the aforementioned, the economic evaluation and the consequent valuation
of a mega-project is very important.[2]

4.2   Ex-Ante Evaluation: The Cost-Benefit Analysis

A valuation method frequently used by business and government officials to support
the ex-ante evaluation is Cost-Benefit Analysis (*CBA*), which, unlike other methods
such as Cost Effectiveness Analysis (*CEA*) or Cost Utility Analysis (*CUA*),[3] provides
a monetary value[4] quantifying the benefits and costs and thus allows a comparison
of wider range of scenarios [27]. Note that, in order to use *CBA*, certain parameters
need to be known or, in absence of real data, estimated[5] [27].

CBA Guide Team [5] and Sartori et al. [26] wrote a guide of methodology for *CBA*
with consideration for the *EU* policies. *CBA* requires an investigation of a project's
net impact on economic welfare, and this is done in five steps:

1. observed prices or public tariffs are converted into shadow prices, which better
   reflect the social opportunity cost of the good;
2. externalities are taken into account by giving them a monetary value;
3. indirect effects (i.e., the effects not already captured by shadow prices) are
   included if relevant;
4. costs and benefits are discounted with a real social discount rate;
5. evaluation is based on economic performance or profitability indicators, i.e. the
   net present value, the profitability index, and the internal rate of return.

The net present value (NPV) is the difference between the present value of cash
inflows and the present value of cash outflows over a period of time $t = 0, \ldots, T$:

---

[2]Whereas the evaluation is intended to be the process of determining the worth of something, the
valuation is the estimation of something's monetary worth.

[3]Whereas *CEA* is a form of economic analysis that compares the relative costs and outcomes
(effects) of different courses of action, *CUA* is an economic analysis used to evaluate medical-
health investment projects in which the incremental cost of a program from a particular point of
view is compared to the incremental health improvement expressed in the unit of quality adjusted
life years (*QALYs*).

[4]Because sometimes it is impossible to give a monetary value to the costs or to the benefits, Guess
and Farnham [17] suggest to use the Risk Analysis (*RA*) method, which "consists of the repeated
random extraction of a set of values for the critical variables, taken within the respective defined
intervals, and then calculating the performance indices for the project resulting from each set of
extracted values" (see [5], p. 63).

[5]Methods frequently used to estimate these input parameters are Scenario Analysis (*SA*), Decision
Trees (*DT*), and Monte Carlo Simulation (*MCS*).

$$NPV = \sum_{t=0}^{T} \frac{IF_t - OF_t}{(1+r)^t} = \sum_{t=1}^{T} \frac{CF_t}{(1+r)^t} - I_0 = PV - I_0 \qquad (4.1)$$

with IF_t the positive cash inflows during the period t, OF_t the negative cash outflows, CF_t the cash flows (that is the net cash inflow-outflows), PV the present value of the future cash flows (i.e., the cash flows from $t = 1$ onwards), I_0 the initial investment (i.e., the negative cash outflow in $t = 0$), and r the discounted rate of return that could be earned in alternative investments. The possible evaluations based on the NPV are *(i)* the acceptance criterion:

$$NPV > 0 \Rightarrow \quad \text{accept,}$$
$$NPV = 0 \Rightarrow \text{indifferent,}$$
$$NPV < 0 \Rightarrow \quad \text{reject,}$$

that is a project with a positive NPV should be accepted,[6] and *(ii)* the preferability criterion:

$$NPV_\alpha > NPV_\beta \Rightarrow \alpha > \beta,$$

which states that the project with the higher NPV should be preferred to another.

The *CBA* is mainly based on the profitability index (PI), which is the ratio of the present value of future cash flows, that is the cash flows starting from $t = 1$, to the initial investment I_0

$$PI = \frac{PV}{I_0} = \frac{\sum_{t=1}^{T} \frac{CF_t}{(1+r)^t}}{I_0} = \frac{\sum_{t=0}^{T} \frac{CF_t}{(1+r)^t} + I_0}{I_0} = \frac{NPV + I_0}{I_0}. \qquad (4.2)$$

The manipulation of the formula (4.2) allows to appreciate how the PI and the NPV derived in (4.1) are closely related to each other. Also on the base of the PI the possible evaluations are *(i)* the acceptance criterion:

$$PI > 1 \Rightarrow \quad \text{accept,}$$
$$PI = 1 \Rightarrow \text{indifferent,}$$
$$PI < 1 \Rightarrow \quad \text{reject,}$$

that is a project with a PI greater than 1 should be accepted,[7] and *(ii)* the preferability criterion:

---

[6]Whereas if a project's NPV is positive or negative, then the project is expected to result in a net gain or loss respectively, if a project's NPV is zero then the project is not expected to result in any significant gain or loss, and thus the investment decision has to be based on non-monetary factors.

[7]Following the same logic used for the NPV, whereas if a project's PI is greater or smaller than 1, then the project is expected to result in a net gain or loss respectively, if a project's PI is equal to 1 then the project is not expected to result in any significant loss or gain, and thus the investment decision has to be based on non-monetary factors.

Table 4.1 Numerical
example of NPV and PI

	α	β	γ
PV	150	100	100
I_0	50	50	20
NPV	100	50	80
PI	3	2	5

$$PI_\alpha > PI_\beta \Rightarrow \alpha \succ \beta,$$

which states that the project with the higher PI should be preferred to another.

However, as stated by the numerical example reported in Table 4.1, whereas in the case of acceptance criterion the NPV and PI provide always the same result, in the case of preferability criterion a conflict between the NPV and PI rankings may occur if the initial investment I_0 is not the same.

In fact, with the same initial investment $I_{0,\alpha} = I_{0,\beta}$ the preferability criteria assert that $NPV_\alpha > NPV_\beta$ and $PI_\alpha > PI_\beta$, but with a different initial investment $I_{0,\alpha} \neq I_{0,\gamma}$ the preferability criteria assert that $NPV_\alpha > NPV_\gamma$ and $PI_\alpha < PI_\gamma$. In the latter case of mutually exclusive decisions, the NPV method should be preferred because the project guaranteeing the largest positive NPV should always be selected.

As shown in Fig. 4.1 the internal rate of return (IRR) is the discount rate that makes the NPV equal zero.

Therefore, as for the PI, also the computation of the IRR relies on the same formula of the NPV

$$NPV = 0 \Rightarrow \sum_{t=0}^{T} \frac{CF_t}{(1+IRR)^t} = 0 \qquad (4.3)$$

(i) The acceptance criterion based on *IRR* asserts that a project should be accepted if the *IRR* is greater than the cost of capital k and rejected if lower:

Fig. 4.1 From NPV to
IRR

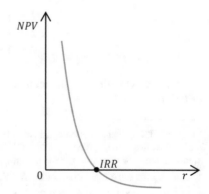

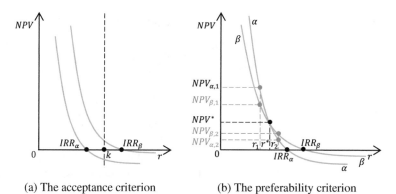

(a) The acceptance criterion   (b) The preferability criterion

Fig. 4.2 The criteria based on IRR

$$IRR > k \Rightarrow \quad \text{accept,}$$
$$IRR = k \Rightarrow \text{indifferent,}$$
$$IRR < k \Rightarrow \quad \text{reject}$$

as shown in panel (a) of Fig. 4.2.

(ii) The preferability criterion based on the *IRR* states that the project with the highest *IRR* not necessarily has to be chosen. Figure 4.2(b) plots the profiles of the two projects α and β, whose characteristics are that the profile of project β is flatter than the profile of project α and thus $IRR_\beta > IRR_\alpha$ and depicts the point r^*, that is the discount rate such that the *NPV* of the two projects to be evaluated are equal $NPV_\alpha = NPV_\beta = NPV^*$: if the discount rate r is such that $r > r^*$ (i.e., at the right of r^*) then $\beta \succ \alpha$ (because $NPV_\beta > NPV_\alpha$) and therefore the project with the greatest *IRR* should be chosen, but if $r < r^*$ (at the left of r^*) then $\alpha \succ \beta$ (because $NPV_\alpha > NPV_\beta$) and therefore the project with the smallest *IRR* should be chosen [3].

Other profitability methods useful with respect to the preferability criterion are the equivalent annuity cash flow methods, which are used to represent a *NPV* as a series of equal cash flows for the length of the projects. (1) The average annual return (*AAR*) is computed as:

$$NPV = \sum_{t=0}^{T} \frac{CF_t}{(1+r)^t} = \sum_{t=1}^{T} \frac{AAR}{(1+r)^t} = \frac{(1+r)^T - 1}{r \cdot (1+r)^T} \cdot AAR$$

$$\Rightarrow AAR = \frac{r \cdot (1+r)^T}{(1+r)^{T+1} - 1} \cdot NPV \tag{4.4}$$

and it is useful for comparing two projects with different length: a project with a greater *AAR* should be preferred. (2) Reminding that $CF_t = IF_t - OF_t = R_t - C_t - \tau \cdot (R_t - C_t)$, with R_t the returns during the period t, C_t the costs, and τ the (constant) tax rate, the equivalent annual cost (EAC) is computed as:

$$\sum_{t=0}^{T} \frac{C_t}{(1+r)^t} - \frac{R_T}{(1+r)^T} = \sum_{t=1}^{T} \frac{EAC}{(1+r)^t} - \frac{(1+r)^T - 1}{r \cdot (1+r)^T} \cdot EAC$$

$$\Rightarrow EAC = \frac{r \cdot (1+r)^T}{(1+r)^T - 1} \cdot \left(\sum_{t=0}^{T} \frac{C_t}{(1+r)^t} - \frac{R_T}{(1+r)^T} \right), \tag{4.5}$$

where R_T is the residual value of the project in T, and it is useful for comparing two projects with the same annual returns, but with different annual costs and/or length. Unlike the AAR, obviously a project with a lower EAC should be preferred.

Besides the five profitability methods described above, the evaluation of a project can also be based on the liquidity method. The pay-back time is a liquidity method and refers to the time required to recoup the funds expended in a project:

$$\sum_{t=0}^{PBT} CF_t = 0 \tag{4.6}$$

If the pay-back time is high then the liquidity reduction extends over long period.[8]

4.3 Ex-Post Evaluation

Whereas the project evaluation using CBA is often done ex-ante, systematic ex-post evaluations are only done infrequently [7]. Nevertheless, ex-post evaluation can be very useful to *check* whether projects really delivered the benefits expected from them at the time, and to *learn* which projects do better and which do worse than expected, and why [7].

In this regards, whereas generally the assessment methods rely on an ex-ante attitude making predictions of how a project might perform, Worsley [30] mentioned that ex-post evaluation, based on the outcomes of past decisions, can serve multiple purposes of which the two primary ones are:

1. accountability and/or control,
2. learning and improvement.

The ex-post evaluation can obviously be carried on the basis of the CBA methodology. Flybjerg et al. [9–11] and Flybjerg [12] are considered key references on the comparison between ex-ante and ex-post evaluation both based on CBA.

---

[8]The discounted pay-back time is computed as $NPV = \sum_{t=0}^{T} CF_t/(1+r)^t \Rightarrow \sum_{t=0}^{PBT} CF_t/(1+r)^t = 0$; note that the discounting lengthens the recovery period.

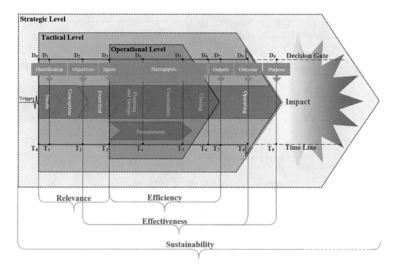

Fig. 4.3 The *PESTOL* model [31, 32]

However, Zidane et al. [31, 32] develop a holistic model for Project Evaluation on Strategic, Tactical and Operational Levels (*PESTOL*) by reviewing different definitions of project success and/or failure and combining the findings with the logic framework (see Fig. 4.3).

The model displayed in Fig. 4.3 has been built by using circular interplay between the logic model and the project life cycle, respectively showed in panels (a) and (b) of Fig. 4.4.

In other words, Zidane et al. [31, 32] initially tried to extract a rational generic project life cycle and thereafter to define a project life cycle that met the logic model.

The evaluation criteria used in the *PESTOL* model by Zidane et al. [31, 32] are the usual ones, that is relevance, effectiveness, efficiency, sustainability, and impact.

4.4 Conclusion

Two of the main causes of failures in mega-project delivery are the ex-ante optimism bias [13] and the lack of post-failure reviews [23], which are closely related to the evaluation process and thus also to the valuation methods. Therefore the research focused on the understanding what causes the many failures in mega-project delivery, and on how to avoid them, must also consider the improvement of evaluation process, also through the perfection of valuation methods, not only ex-ante but also ex-post.

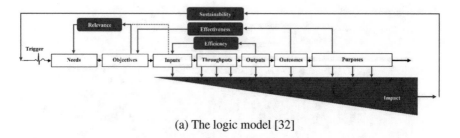

(a) The logic model [32]

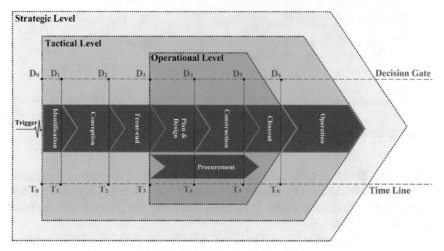

(b) The project life cycle model [32]

Fig. 4.4 The derivation of the *PESTOL* model [32]

References

1. Atkin B (2015) Megaprojects planning and management: essential readings. Constr Manage Econ 33(9):771–774
2. Atkinson R (1999) Project management: cost, time and quality, two best guesses and a phenomenon, its time to accept other success criteria. Int J Project Manage 17(6):337–342
3. Belyadi H, Fathi E, Belyadi F (2017) Economic evaluation. In: Belyadi H, Fathi E, Belyadi F (ed) Hydraulic fracturing in unconventional reservoirs, Gulf Professional Publishing—Elsevier, Amsterdam, pp 324–392. Chapter 18
4. Bovensiepen J, Meitzner Yoder LS (2018) Introduction: the political dynamics and social effects of megaproject development. Asia Pac J Anthropol 19(5):381–394
5. CBA Guide Team (2008) Guide to cost-benefit analysis of investment projects—structural funds, cohesion fund and instrumental for pre-accession. European Commission—Directorate-General for Regional and Urban Policy—Publication Office, Brussels
6. Cleland, DI, King WR (1983) Project management handbook. Wiley Online Library—Wiley, Chichester
7. de Jong G, Vignetti S, Pancotti C (2018) Ex post evaluation of major transport infrastructure projects. In: European transport conference 2018, Association for European transport (AET) 2018 and contributors

8. Flyvbjerg B, Bruzelius N, Rothengatter W (2003) Megaprojects and risk: an anatomy of ambition. Cambridge University Press, Cambridge
9. Flybjerg B, Holm MKS, Buhl SL (2003) How common and how large are cost overruns in transport infrastructure projects? Transp Rev 23(1):71–88
10. Flybjerg B, Holm MKS, Buhl SL (2004) What causes cost overrun in transport infrastructure projects? Transp Rev 24(1):3–18
11. Flybjerg B, Holm MKS, Buhl SL (2005) How (in)accurate are demand forecasts in public works projects? The case of transportation. J Am Plan Assoc 71(2):131–146
12. Flybjerg B (2007) Policy and planning for large-infrastructure projects: problems, causes, cures. Environ Plan B 34(4):578–597
13. Flyvbjerg B (2010) Over budget, over time, over and over again—managing major projects. In: Morris WG, Pinto JK, Söderlund J (ed) The Oxford handbook of project management. Oxford University Press, Oxford, pp 321–344. Chapter 13
14. Flyvbjerg B (2014) What you should know about megaprojects and why: an overview. Proj Manag J 45(2):6–19
15. Flyvbjerg B (2017) Introduction: the iron law of megaproject management. In: Flyvbjerg B (ed) The Oxford handbook of megaproject management. Oxford University Press, Oxford, pp 1–18. Chapter 1
16. Garemo N, Matzinger S, Palter R (2015) Megaprojects: the good, the bad, and the better. McKinsey & Company—Capital Projects & Infrastructure, July
17. Guess GM, Farnham PG (2000) Cases in public policy analysis. Georgetown University Press, Washington DC
18. Ika LA, Diallo A, Thuillier D (2012) Critical success factors for World Bank projects: an empirical investigation. J Proj Manag 30(1):105–116
19. JICA (2004) Guideline for project evaluation—practical methods for project evaluation. Office of Evaluation, Planning and Coordination Department—Japan International Cooperation Agency (JICA)
20. Mišić S, Radujković M (2015) Critical drivers of megaprojects success and failure. Procedia Eng 122(9):71–80
21. Odeck J (2004) Cost overruns in road construction–what are their sizes and determinants? Transp Policy 11(1):43–53
22. OECD (2009) Glossary of key terms in evaluation and results based management. OECD Publications, Paris
23. Pinto JK, Kharbanda OP (1996) How to fail in project management (without really trying). Bus Oriz 39(4):45–53
24. Pinto J, Mantel S (1990) The cause of project failure. IEEE Trans Eng Manage 37(4):269–276
25. Samset K (2003) Project evaluation—making investment succeed. Fagbokforlaget, Bergen
26. Sartori D, Catalano G, Genco M, Pancotti C, Sirtori E, Del Bo C (2014) Guide to cost-benefit analysis of investment projects—economic appraisal tool for cohesion policy 2014–2020. European Commission—Directorate-General for Regional and Urban Policy—Publication Office, Brussels
27. Sherman G, Siebers P-O, Menachof D, Aickelin U (2012) Evaluating different cost-benefit analysis methods for port security operations. In: Faulin J, Juan AA, Grasman SE, Fry MJ (ed) Decision making in service industries: a practical approach. CRC Press—Taylor & Francis Group, Boca Raton, pp 279–302. Chapter 12
28. Tabish SZS, Jha KN (2011) Identification and evaluation of success factors for public construction projects. Constr Manag Econ 29(8):809–823
29. van Wee B, Tavasszy LA (2008) Ex-ante evaluation of mega-projects: methodological issues and cost-benefit analysis. In: Priemus H, Flyvbjerg B, van Wee B (ed) Decision-making on mega-projects—cost-benefit analysis, planning and innovation, pp 40–65, Chapter 3
30. Worsley T (2014) Ex-post assessment of transport investments and policy interventions: roundtable summary and conclusions. International Transport Forum Discussion Papers, No. 19, OECD Publications, Paris

31. Zidane YJ-T, Johansen A, Ekambarum A (2015) Project evaluation holistic framework—application on megaproject case. Procedia Comput Sci 64:409–416
32. Zidane YJ-T, Johansen A, Husseun BA, Andersen B (2016) *PESTOL*—framework for project evaluation on strategic, tactical and operational levels. Int J Inf Syst Proj Manag 4(3):25–41

Chapter 5
Using Real Options to Value Two Key Merits of Small Modular Reactors

Giorgio Locatelli[ID], Marco Pecoraro, Giovanni Meroni, and Mauro Mancini

Abstract Small Modular Reactors (SMRs) are a better choice respect to large reactors, according to their inherent construction flexibility and their short construction time. In the energy sector, the Discounted Cash Flow (DCF) is usually used to evaluate an investment. However, "Real Options" method is a better choice respect to DCF approach underestimates the importance of flexibility during the investment decision. In this chapter, starting from real options approach, two fundamental characteristics for the choice of the construction of SMRs are defined: the time to market and the adding of a new plant on a typical portfolio. The computational model described in this chapter assesses the superior performance of SMRs in the UK.

Keywords Energy · Economics and finance · Management

5.1 Introduction

The International Atomic Energy Agency [1] defines Small Modular Reactors (SMRs) as *"newer generation reactors designed to generate electric power up to 300 MW, whose components and systems can be shop fabricated and then transported as modules to the sites for installation as demand arises"*. Several SMRs designs, detailed in [1–3], are currently at different stages of development. Reference [4] provides a good summary of the innovative feature of SMRs; *"reactor designs that are deliberately small, i.e. designs that do not scale to large sizes but rather capitalize on their smallness to achieve specific performance characteristics."*

A recent review related to the economics and finance of SMR is presented in [5].

One of the key SMR characteristics is the smaller size with respect to traditional Large Reactors (LRs), determining the lack of the economy of scale [6, 7]. Several documents discuss the competitiveness of SMRs versus LRs and how SMRs might

G. Locatelli (✉)
University of Leeds, Leeds LS2 9JT, UK
e-mail: g.locatelli@leeds.ac.uk

M. Pecoraro · G. Meroni · M. Mancini
Politecnico di Milano, Milan 20156, Italy

55

balance the "diseconomy of scale" with the "economy of multiples" [8–10]. Modularisation (factory fabrication of modules, transportation and installation on-site [11]) allows working in a better-controlled environment reducing cost and schedule [7, 12–15]. The degree of modularisation influences SMR capital costs [13, 16]. Furthermore, SMR smaller size and simpler design further reduce construction schedule [9, 17]. The SMR expected schedule is 4/5 years for the FOAK (first-of-a-kind) and 3/4 years for the NOAK (nth-of-a-kind), even if licensing can be a challenge [18]. SMR incremental capacity addition can allow overcoming one of the key barriers in building LRs: the massive and risky upfront investment [19].

Another key SMR advantage to consider is the possibility to have multiples units on the same site [8, 12, 20]. Other factors to consider in SMR economic and financial evaluation are: suitability for cogeneration [21–24], expected higher learning rate [10, 25], better adaptability to market [8], equal or higher capacity factor than current LRs [26]. Once these factors are taken into account, the capital cost is comparable between the two technologies [27].

One of the key SMRs advantages is the possibility to split a large investment into smaller ones. The construction of a single LR is a risky investment. The construction of n SMRs is an investment decision with n degrees of freedom that allows hedging investment risks. The economic merit of flexibility can be calculated using the Real Options (ROs) approach.

The construction of large projects in general, and Power Plants (PPs) in particular, is jeopardised by over budget and delay [28–30] Since ROs assess the decision maker's options (i.e. degrees of freedom) to hedge the investment risks, they increase the expected returns and, at the same time, minimise the volatility of the investment considered. *"RO theory postulates that projects under uncertainty might possess RO; the projects become flexible if the RO can be identified and timely executed; flexibility adds value to the projects"* [31]. In the energy sector the RO model evaluates opportunities such as waiting for the most advantageous moment to invest; abandon a not profitable investment; switching from a technology to a more profitable one, produce outputs for more than one market [32] etc.

This chapter presents an appraisal based on ROs with two key peculiarities:

1. The modelling of the Time to Market (TTM) effect.
2. The investment in a certain Power Plant (PP) considering the utility portfolio.

The TTM is the time from a product concept definition to its availability for sale [33]. In this chapter, the TTM is the time between the decision to build a PP and the beginning of commercial operations. In the energy sector reducing the TTM means reducing the risk (e.g. from electricity price fluctuation) and collecting early revenues increasing the Net Present Value (NPV). SMRs can be built faster than LRs, which is a relevant aspect to consider. The utility portfolio is relevant since PPs might not be considered as a "single asset" investment but must fit into a broader strategy of a utility owning a portfolio of PPs.

5.2 Real Options in the Energy Sector

Real options model has been applied in the energy sector by a lot of researchers since the 70s. Reference [34] introduce for the first time the term "Real Option", and he highlighted the possibility to consider an investment as a call option on a real asset. Reference [35] underlined the key role of RO in the evaluation of investments within an uncertain scenario. Even [36] focussed on RO advantage to support the decision making process in an uncertain surrounding business. RO method allows to establish the optimal project in terms of cash flow and to consider the better choice of investment in terms of the exercise rules of the option [37].

It is possible to highlight three important RO in previous literature [38]: invest, defer, abandon. Table 5.1 summarises relevant examples of papers applying the RO theory in the power and energy sector and shows the difference with the RO approach explained in this chapter.

It is also possible to join in a "compound option" two or more ROs to practically evaluate a more realistic scenario. For example, it is possible to create the "wait and build" option that is the compound option joining the option to build and the option to wait. This compound option allows introducing in the model the possibility for the investor to realize a postponed investment.

There are some examples of compound option in literature, also in other sectors rather than energy and power. Reference [39] applies RO method to evaluate the risky coupon bond problem considering how to price an option on another one that is a type of compound options approach. Reference [40] apply the RO method to renewable energy power plants life-cycle. Reference [41] introduce an evaluation of compound options using a binomial lattice model to create a versatile management framework. Reference [42] use compound option to evaluate different possibilities for energy storage. In this chapter, the compound options method is used to simulate the typical "Stage-gate process" of the energy sector.

5.3 Portfolio Analysis

The boundary conditions of the electricity markets encourage utilities to diversify their portfolio. Reference [49] highlights that a new investment has always to consider the company portfolio: "*The risk position of the company is determined by the entire portfolio and the interaction of various positions. Therefore, the decision to enter into new contracts cannot be taken independently from the current portfolio*". Portfolio management was initially stated for the financial sector and then converted for the energetic company uses. The model proposed in this chapter is a development of the Mean-Variance Portfolio theory (MVP) because:

- It is a well-known method, used a lot in literature and suitable for decision-makers;
- It is versatile and it can be used with different objectives (e.g. maximisation of the NPV Mean; minimisation of risk);

Table 5.1 Recent applications of real options in the energy field

	This work	[44]	[45]	[46]	[47]	[48]
Scope of work	Building a realistic investment model in the energy sector by considering the TTM Effect and the Actual Portfolio of a utility.	Evaluate how all risks and uncertainties impact on the development of new Nuclear PP in China	Help a utility determine the value of sequential SMR	Analyse different RO model type to assess the best for making an investment in the energy sector.	Examine energy switching from non-renewable to renewable technologies in Mongolia	Apply a RO to a mini-hydro PP case comparing its results with the results obtainable through the classical DCF approach
Real option evaluation method	Simulation with optimized exercise thresholds (see [43])	Partial differential equation	Dynamic programming method	Binomial tree	Simulation method	Binomial tree
Options considered	Compound option; option to invest/abandon/defer/choose	Compound option; option to invest/abandon	Option to invest; option to abandon	Option to defer; option to invest	Option to switch	Option to invest
Outputs	E(NPV); σ(NPV); exercise thresholds; efficient frontier 2D for each technology; efficient frontier 3D for each portfolio	Value of the option	E(NPV); classical NPV	Value of the options	Decision to switch; option value	E(NPV); classical NPV; option value

NPV net present value, *E* expected, *σ* standard deviation [43]

- It can be used to evaluate a portfolio of investment or a single investment joining also RO method;
- Every single portfolio can be directly compared to the others in terms of Expected NPV (E(NPV)) and risk (σ(NPV)).

Reference [50] substantially starts research about MVP. According to the MVP theory presented in [51], each portfolio has two attributes: its mean value (μ) and its standard deviation (σ). The mean value is the mean value of the controlled variables (e.g. the NPV), while the σ represents the risk on the investment. In the energy sector, it is possible to obtain a lot of different portfolios, combining different type and numbers of power plants. Each portfolio is characterised by its μ and σ. However, only few of them represent a rational choice because, given a certain μ, it is reasonable to choose only the portfolio with the lowest σ, i.e., the lowest risk. Alternatively, from the opposite point of view, given a certain σ, a reasonable investor implements only the portfolio with the highest μ; therefore, there is a one-to-one link among μ and σ. Given a certain level of μ, the more beneficial σ is automatically linked (and vice versa). Considering Fig. 5.1, it is possible to highlight the so-called "efficient frontier", the continuous line from "A" to "B", where all the best portfolios are. "A" is the portfolio with the lowest return and risk, while "B" has the highest return and risk. "C" is another optimal portfolio because, given a certain level of risk, it maximises the return or, given a certain level of return, it minimises the risk. "D" is not a rationale portfolio since, for the same risk, the "C" portfolio provides a higher

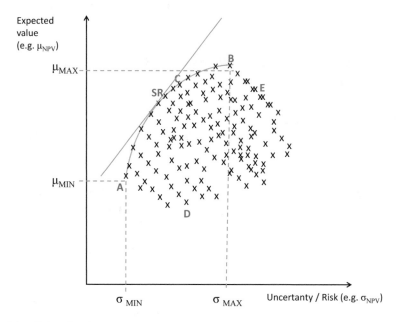

Fig. 5.1 Efficient frontier: the classical MVP theory [43]

return. "E" is not a rationale portfolio since, for the same expected return, the C portfolio has the lowest risk.

The efficient frontier collects all the optimal portfolios [52] introduces a parameter, the *Sharpe Ratio* (SR), that compares the optimal portfolios considering their return over risk. According to [53] the SR is a measure for calculating the risk-adjusted return, and this ratio has become the industry standard for such calculations. The SR is the average return earned in excess of the risk-free rate per unit of volatility or total risk. Subtracting the risk-free rate from the mean return (the expected value of all the likely returns of investments comprising a portfolio), the performance associated with risk-taking activities can be isolated. One intuition of this calculation is that a portfolio engaging in "zero risk" investment, such as the purchase of U.S. Treasury bills (for which the expected return is the risk-free rate), has a SR equal to zero. Generally, the greater the value of the SR, the more attractive the risk-adjusted return. The investor is likely to prefer the portfolio on the efficient frontier with the highest expected return for the unit of risk (i.e. the highest SR). Geometrically the point of the efficient frontier that corresponds to the solution of this problem is tangent to the efficient frontier: the optimal portfolio received is called "Tangent Portfolio".

This work overcomes one of the principal drawbacks in the literature about the MVP that limits its use: "MVP is a static methodology, heavily relying on past data. As a result, a portfolio that is thought of as optimal today might already be way off the efficient frontier tomorrow, depending on how the environment has changed. It is therefore a method that should only be considered within a very limited time frame" [54]. The application of RO to the portfolio analysis tackles this limitation.

5.3.1 Application of Real Options to Perform Portfolio Analysis

In the literature, there are only a few examples of the application of RO to perform a portfolio analysis. Table 5.2 benchmark this work with respect to the literature.

5.4 According to Table 5.2, the Steps Followed by Classical Approaches to Perform a Portfolio Analysis Are

1. Consider the historical data to identify the best strategy of investment for a certain plant type.
2. Calculate the E(NPV) and σ(NPV) of the overall portfolio.
3. Compare all the possible portfolios through the efficient frontier method to identify the one maximising the profit for a specific level of risk.

The computational model presented on the following is based on a set of different exercise thresholds (see [43] for a discussion on exercise thresholds) and it calculates

Table 5.2 Examples of real options application to perform portfolio analysis [43]

	This work	[45]	[55]	[56]
RO evaluation method	Simulation with optimized exercise thresholds	Stochastic grid bundling method	Partial differential equations	Dynamic programming method
Options considered	Compound options; option to invest; option to choose; option to abandon	Option to invest; option to abandon	Respectively option to invest and to abandon	Option to invest
TTM effect	Modelled	Not considered	Not considered	Not considered
Pre-operating phases	Modelled as the succession of three compound options	Only the construction phase is considered	Only the construction phase is considered	Only the construction phase is considered
Actual portfolio	Influence results	Results not influenced	Influence results	Results not influenced
Method used to perform the portfolio analysis	MVP theory	MVP theory	Stochastic dominance	CVaR method
OUTPUT indicators	E(NPV); σ(NPV); exercise thresholds; efficient frontier 2D for each technology; efficient frontier 3D for portfolio	Efficient frontier 2D for portfolio in which every technology is a single static point on it; value of the option	Value of the option. The efficient frontier is not built: the PDE do not find out the level of risk of the investment	Expected cost; level of risk; a single technology is a single static point on the plane E(cost)—level of risk

the different effects on the output distribution of the overall portfolio, starting from these exercise thresholds. Therefore, each portfolio in the Cartesian graph E(NPV)—σ(NPV) is a function of the values of the exercise thresholds that, triggering the options in different conditions, modify the NPV distribution of the overall portfolio. Unfortunately, RO analysis of a real portfolio of power plants is very complex, and its application in a real situation is often impossible. However, the method described in the next sections overcomes this problem because it guarantees to the model users to analyse cases of investment in real portfolios with roughly the same effort of investments in simple portfolios.

5.5   Application of RO to a Hypothetical Portfolio

The main results are summarised in Table 5.3. The starting point for the computational simulation is the necessity for the utility to satisfy an increased demand for electricity, equal to 1.5 GW, within 20 years. Therefore, the RO method is used to evaluate the efficient frontier for the utility portfolio as a function of the exercise thresholds. A synthesis of the main results is supplied by Fig. 5.2 and Table 5.4; however, it is possible to highlight:

- The option has to be exercised when $P_{threshold} > P_0$ (i.e. it is worth waiting)
- After a specific value P_{lim} the points do not belong on the efficient frontier anymore (i.e. the decision has to be taken in a finite time)

Table 5.3 Composition of the hypothetical existing portfolio

Technology	Capacity installed [MW]	% in the overall actual portfolio
Nuclear	1500	46.15%
Coal	750	23.08%
CCGT	1000	30.77%

CCGT combined cycle gas turbine [43]

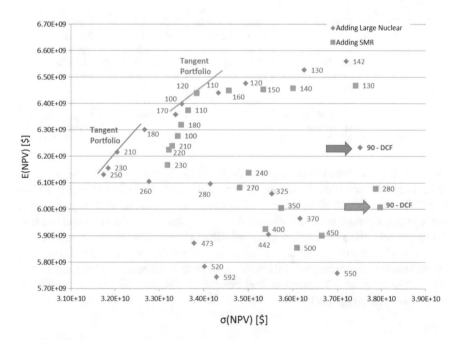

Fig. 5.2 Efficient frontier of the portfolio with an additional LR or equivalent SMR power [43]

Table 5.4 Results obtained with the hypothetical portfolio [43]

Additional PP	Lower bound efficient frontier	Upper bound efficient frontier	Tangent portfolio condition
Large nuclear	$P_{LB}^* = 100\$/MWh$	$P_{UB}^* = 250\$/MWh$	$P_{SR}^* = 210\$/MWh$
SMR	$P_{LB}^* = 100\$/MWh$	$P_{UB}^* = 230\$/MWh$	$P_{SR}^* = 120\$/MWh$

- The standard DCF method supplies a useless static evaluation of the efficient frontier respect to the RO method that creates a lot of possible dynamic scenarios identifying the best options with an optimised efficient frontier.
- The points on the efficient frontier have these properties:

  - The option has to be exercised when $P_{threshold} > P_0$ (i.e. it is worth waiting)
  - After a specific value P_{lim} the points do not belong on the efficient frontier anymore (i.e. the decision has to be taken in a finite time).

- All the points on the efficient frontier have these characteristics:

  - The condition to find them is to exercise the option only when $P_0 < P_{threshold} \leq P_{lim}$
  - The portfolio on the efficient frontier can be compared in terms of SR. This is a value of the exercise threshold P_{SR} that corresponds to the tangent portfolio of the efficient frontier.

The RO method allows a more complete evaluation of the investment's strategy than the static DCF method; RO model introduces the optimised efficient frontier, as it is shown by Fig. 5.2. The optimised efficient frontier is composed of the best investment strategy as a function of the exercise thresholds and the desired level of risk. Table 5.5 shows how the decision to invest varies according to the specific objective function. The decision-maker can choose the most suitable PP depending on his/her risk appetite.

Table 5.5 Improvement of results guaranteed by this method [43]

Objective function	Large reactor's results	SMR's results	PP chosen	Condition of investment
Maximization of NPV mean	E(NPV) = 6560 Mln$	E(NPV) = 37,193 Mln$	Large reactor	$P^* =$ 142$/MWh
Minimization of σ NPV	E(NPV) = 6131 Mln$	σ(NPV) = 31,786 Mln$	Large reactor	$P^* =$ 250$/MWh
Maximization of the SR value	SR = 0.191	SR = 0.197	SMR	$P^* =$ 120$/MWh

Table 5.6 The deterministic inputs used in this work [43]

	Nuclear	Coal	CCGT	SMR
Capacity [MW]	1500	750	500	335
Capacity factor (%)	85	85	85	95
Overnights cost [$/KW]	5335	3220	1003	6362
O&M Cost [$/MWh]	13.96	13.4	15.03	21.28
Fuel cost [$/MWh]	8.26	22.27	47.4	8.26
Carbon cost [$/MWh]	0	23.96	10.54	0
Construction time [years]	6	4	3	5
Study time [years]	1	/	/	1
Design time [years]	2	/	/	2
Life [years]	60	40	30	60

5.5.1 Application to a UK Utility Portfolio

In this section the application of RO model to a UK utility portfolio is described, the model is built using both deterministic and stochastic input data, as it is shown by Table 5.6. The stochastic variables have been modelled as Geometric Brownian Motion (GBM). Consistently with [57] the initial values of the variables modelled with the GBM model are: gas cost 47.39 $/MWh; coal cost 22.27 $/MWh and for the electricity price a value of 90 $/MWh.

The GBM functional form applied in this work on the electricity price is:

$$dp = \alpha p dt + \sigma p dz \tag{5.1}$$

where α is the drift, σ is the volatility of the process in a time period, t is the time, and z is a Wiener process, where Wiener process is a continuous-time Gaussian process with independent increments used for modelling the Brownian motion. This model does not consider the Mean Reversion, Spike Jumps and Price Proportional Volatility.

The main advantages of the proposed model are:

1. The computational simplicity. The model is based on a time-independent Monte Carlo simulation and it considers only the price and the volatility of the electricity price at time zero. However, to consider also the long-term uncertainty the electricity price is evaluated through a GBM process:

$$E(P_{t+n}) = P_t \tag{5.2}$$

$$Var(P_{t+n}) = P_t^2 \left(e^{\sigma^2} n - 1 \right) \tag{5.3}$$

2. It can be implemented in a simple Excel spreadsheet.
3. The model is coherent even with the removal of the mean reversion and jumps because it considers a baseload scenario and so the importance of these two parameters is very low.

The following equation models the electricity price:

$$P_{t+1} = P_t + \sigma P_t W_t \tag{5.4}$$

where P_t is the electricity price at time t, W_t is a standard normal variable, σ is the volatility of the process in a time period and $\sigma = 0.3$ as in [51].

The same description can be considered for the gas cost and for the coal cost too.

The Total Capital Investment Cost (TCIC) follows the method developed by [35] and [58] as:

$$dK = -Idt + \sigma(IK)^{1/2}dz \tag{5.5}$$

As in [58] the mean and the variance of the TCIC are described by these relationships:

$$E(TCIC) = K \tag{5.6}$$

$$Var(TCIC) = \frac{\sigma^2 K^2}{2 - \sigma} \tag{5.7}$$

Since the evaluation model is discrete time based this stochastic process is modelled with:

$$K_{t+1} = K_t - I + \sigma(IK)^{1/2}W_t \tag{5.8}$$

where W_t is a standard normal variable that gives the variability to this data.

The computational model considers a speculated portfolio of a typical UK energy utility and it is composed by the following contributions in terms of electric power output: 116 MWe from renewable power plants, 8741 MWe from nuclear power plants, 3987 MWe from coal power plants and 1306 MWe from gas power plants.

Figure 5.3 summaries the results of the model described in [43]:

1. Without considering the compound options in the pre-operational phase, every PP portfolio solution belongs to the efficient frontier. In this case, an investment in SMR or LR supplies a profit greater than the profit provided by a gas PP or a coal PP but it has a higher level of risk.
2. Considering a flexible approach in the pre-operational phase with the compound option method, the simulation output changes a lot. The risk related to SMR and LR is reduced by the possibility to abandon the investment if it is not profitable anymore.

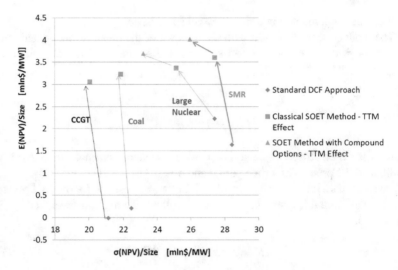

Fig. 5.3 Comparison between the results considering the TTM effect [43]

In Fig. 5.4, a comparison between three of the possible output of the computational model is shown the classical DCF approach with the MVP Theory for each of the four possible additional PP

1. The classical DCF approach with the MVP Theory for each of the possible additional PP;
2. The value of the compound options with the option to defer for each of the possible additional PP;

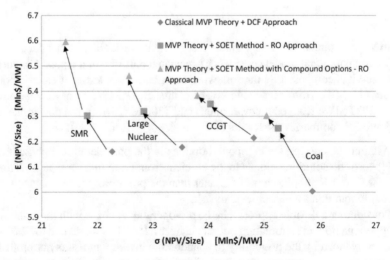

Fig. 5.4 Results obtained considering the utility portfolio in the UK [43]

3. The value of the pre-operational phase of the additional PPs as the succession of three sequential compound options with the option to defer.

Figure 5.4 shows how compound options increase more the value of the investment in SMR or LR than investment in CCGT or Coal PPs.

5.6 Conclusions

The main conclusion of the proposed model is a confirmation of the economic profitability of an investment in SMR, that is independent from an important factor such as the diseconomy of scale. Some of the main advantages of SMRs are the shorter construction times respect to LRs and the possibility to obtain a greater variation of the portfolio according to the smaller size of the investment rather than considering LRs. The advantages of SMRs respect to LRs are mainly related to an intuitive idea of reduction of the risk of the investment, but a key point is to quantitative evaluate these advantages. In this chapter, the issue is tackled through the development of a computational model based on the RO approach to evaluate the investment appraisal in the energy sector. Some of the most important aspects of the proposed model are the inclusion of the Time to Market (TTM) evaluation and the assessment of the overall "portfolio effect".

TTM evaluation is very useful because it allows to consider the economic value of longer construction time, such as for a nuclear power plant respect to a conventional plant, and, at the same time, to assess the possibility to abandon the project if the market conditions change. Even if the MVP is a traditional static method (very dependent from past data), in this model the portfolio analysis is dynamic, thanks to RO approach that allows the assessment of an optimised efficient frontier with an added degree of freedom respect to traditional models. The RO method allows to consider also the more relevant options in the future, so the resulting model has a dynamic component. The complete evaluation of the result obtained by the model considering a speculated utility portfolio in the UK, suggests the SMRs as ideal option to maximise the NPV mean and the SR minimising the risk respect to the profit.

References

1. IAEA (2016) Advances in small modular reactor technology developments
2. Hidayatullah H, Susyadi S, Subki MH (2015) Design and technology development for small modular reactors—safety expectations, prospects and impediments of their deployment. Prog Nucl Energy 79:127–135. https://doi.org/10.1016/j.pnucene.2014.11.010
3. Karol M, John T, Zhao J (2015) Small and medium sized reactors (SMR): a review of technology. Renew Sustain Energy Rev 44:643–656. https://doi.org/10.1016/j.rser.2015.01.006
4. Ingersoll DT (2009) Deliberately small reactors and the second nuclear era. Prog Nucl Energy 51:589–603

5. Mignacca B, Locatelli G (2020) Economics and finance of small modular reactors: a systematic review and research agenda. Renew Sustain Energy Rev 118. https://doi.org/10.1016/j.rser.2019.109519
6. Lokhov A, Cameron R, Sozoniuk V (2013) OECD/NEA study on the economics and market of small reactors. Nucl Eng Technol 45:701–706. https://doi.org/10.5516/NET.02.2013.517
7. Carelli MD, Ingersoll DT (2014) Handbook of small modular nuclear reactors
8. Barenghi S, Boarin S, Ricotti ME (2012) Investment in different sized SMRs : economic evaluation of stochastic scenarios by INCAS code. In: International congress on advances in nuclear power plants 2012, ICAPP, Chicago, USA, pp 2783–2792
9. Vegel B, Quinn JC (2017) Economic evaluation of small modular nuclear reactors and the complications of regulatory fee structures. Energy Policy 104:395–403
10. EY (2016) Small modular reactors—can building nuclear power become more cost-effective?
11. Mignacca B, Alawneh AH, Locatelli G (2019) Transportation of small modular reactor modules: what do the experts say? In: 27th International conference on nuclear engineering (ICONE27). Japan Society of Mechanical Engineers, Tsukuba, Japan, 19–24 May 2019
12. Boldon L, Sabharwall P, Painter C, Liu L (2014) An overview of small modular reactors: status of global development, potential design advantages, and methods for economic assessment. Int J Energy, Environ Econ 22:437–459
13. Maronati G, Petrovic B, Van Wyk JJ, Kelley MH, White CC (2017) EVAL: a methodological approach to identify NPP total capital investment cost drivers and sensitivities
14. Mignacca B, Locatelli G, Alaassar M, Invernizzi D (2018) We never built small modular reactors (SMRs), but what do we know about modularization in construction? In: 26th International conference on nuclear engineering (ICONE 26). ASME
15. Lloyd CA, Roulstone A (2018) A methodology to determine SMR build schedule and the impact of modularisation. In: 26th International conference on nuclear engineering, ICONE26, Lundon, United Kingdom, pp 1–8
16. Maronati G, Petrovic B, Banner JW, White CC, Kelley MH, Van Wyk J (2016) Total capital investment cost evaluation of SMR modular construction designs. In: International congress on advances in nuclear power plants, ICAPP 2016, pp 943–948
17. Playbell I (2016) Economy, safety and applicability of small modular reactors. In: Proceedings of the institution of civil engineers
18. Sainati T, Locatelli G, Brookes N (2015) Small modular reactors: licensing constraints and the way forward. Energy 82:1092–1095. https://doi.org/10.1016/j.energy.2014.12.079
19. Sainati T, Locatelli G, Smith N (2019) Project financing in nuclear new build, why not? The legal and regulatory barriers. Energy Policy 129:111–119
20. Boarin S, Ricotti ME (2014) An evaluation of SMR economic attractiveness. Sci Technol Nucl Install 2014:1–8. https://doi.org/10.1155/2014/803698
21. Locatelli G, Boarin S, Fiordaliso A, Ricotti M (2017) Load following by cogeneration: options for small modular reactors, gen IV reactor and traditional large plants. In: 25th International conference on nuclear engineering, ICONE 2017; Shanghai; China. American Society of Mechanical Engineers (ASME)
22. Locatelli G, Fiordaliso A, Boarin S, Ricotti ME (2017) Cogeneration: an option to facilitate load following in small modular reactors
23. Ingersoll DT, Houghton ZJ, Bromm R, Desportes C (2014) NuScale small modular reactor for co-generation of electricity and water. Desalination 340:84–93. https://doi.org/10.1016/j.desal.2014.02.023
24. Ingersoll DT, Houghton ZJ, Bromm R, Desportes C (2014) Integration of NuScale SMR with desalination technologies. In: ASME 2014 small modular reactors symposium, SMR 2014, Washington, DC, USA
25. NNL (2014) Small modular reactors (SMRs) feasibility study (2014)
26. Shropshire D (2011) Economic viability of small to medium-sized reactors deployed in future European energy markets. Prog Nucl Energy 53:299–307. https://doi.org/10.1016/j.pnucene.2010.12.004

27. Carelli MD, Mycoff CW, Garrone P, Locatelli G, Mancini M, Ricotti ME, Trianni A, Trucco P (2008) Competitiveness of small-medium, new generation reactors: a comparative study on capital and O&M costs. In: International conference on nuclear engineering, proceedings, ICONE

28. Flyvbjerg B (2006) From Nobel prize to project management: getting risks right. Proj Manag J 37:5–15

29. Invernizzi DC, Locatelli G, Brookes NJ (2018) Cost overruns—helping to define what they really mean. Proc Inst Civ Eng Civ Eng 171:85–90. https://doi.org/10.1680/jcien.17.00001

30. Locatelli G (2018) Why are megaprojects, including nuclear power plants, delivered overbudget and late? Reasons and remedies—report MIT-ANP-TR-172, Center for Advanced Nuclear Energy Systems (CANES), Massachusetts Institute of Technology

31. Martínez Ceseña EA, Mutale J, Rivas-Dávalos F (2013) Real options theory applied to electricity generation projects: a review. Renew Sustain Energy Rev 19:573–581. https://doi.org/10.1016/j.rser.2012.11.059

32. Locatelli G, Boarin S, Pellegrino F, Ricotti ME (2015) Load following with small modular reactors (SMR): a real options analysis. Energy 80:41–54

33. Cohen MA, Eliashberg J, Ho TH (1996) New product development: the performance and time-to-market tradeoff. Manage Sci 42:173–186

34. Myers SC (1977) Determinants of corporate borrowing. J Financ Econ 5:147–175. https://doi.org/10.1016/0304-405X(77)90015-0

35. Pindyck RS (1992) Investments of uncertain cost

36. Driouchi T, Bennett DJ (2012) Real options in management and organizational strategy: a review of decision-making and performance implications. Int J Manag Rev 14:39–62. https://doi.org/10.1111/j.1468-2370.2011.00304.x

37. He Y (2007) Real options in the energy markets

38. Kodukula DP, Papudesu C (2007) Project valuation using real options—a practiotioner's guide

39. Geske R (1977) The valuation of corporate liabilities as compound options. J Financ Quant Anal 12:541–552. https://doi.org/10.2307/2330330

40. Siddiqui AS, Marnay C, Wiser RH (2006) Real options valuation of US federal renewable energy research, demonstration, and deployment. Energy Policy 35:265–279. https://doi.org/10.1016/j.enpol.2005.11.019

41. Cheng C-T, Lo S-L, Lin TT (2011) Applying real options analysis to assess cleaner energy development strategies. Energy Policy 39:5929–5938. https://doi.org/10.1016/j.enpol.2011.06.048

42. Locatelli G, Invernizzi DC, Mancini M (2016) Investment and risk appraisal in energy storage systems: a real options approach. Energy 104:114–131. https://doi.org/10.1016/j.energy.2016.03.098

43. Locatelli G, Pecoraro M, Meroni G, Mancini M (2017) Appraisal of small modular nuclear reactors with 'real options' valuation. Proc Inst Civ Eng—Energy 170:51–66

44. Shi H, Song H (2013) Applying the real option approach on nuclear power project decision making. Energy Procedia 39:193–198. https://doi.org/10.1016/j.egypro.2013.07.206

45. Jain S, Roelofs F, Oosterlee CW (2013) Construction strategies and lifetime uncertainties for nuclear projects: a real option analysis. Nucl Eng Des 265:319–329. https://doi.org/10.1016/j.nucengdes.2013.08.060

46. Zambujal-Oliveira J (2013) Investments in combined cycle natural gas-fired systems: a real options analysis. Int J Electr Power Energy Syst 49:1–7. https://doi.org/10.1016/j.ijepes.2012.11.015

47. Detert N, Kotani K (2013) Real options approach to renewable energy investments in Mongolia. Energy Policy 56:136–150. https://doi.org/10.1016/j.enpol.2012.12.003

48. Santos L, Soares I, Mendes C, Ferreira P (2014) Real options versus traditional methods to assess renewable energy projects. Renew Energy 68:588–594. https://doi.org/10.1016/j.renene.2014.01.038

49. Hlouskova J, Kossmeier S, Obersteiner M, Schnabl A (2005) Real options and the value of generation capacity in the German electricity market. Rev Financ Econ 14:297–310. https://doi.org/10.1016/j.rfe.2004.12.001

50. Markowitz H (1952) Portfolio selection. J Financ 7:77–91. https://doi.org/10.2307/2329297
51. Locatelli G, Mancini M (2011) Large and small baseload power plants: drivers to define the optimal portfolios. Energy Policy 39:7762–7775. https://doi.org/10.1016/j.enpol.2011.09.022
52. Sharpe WF (1994) The sharpe ratio. J Portf Manag 21:49–58. https://doi.org/10.3905/jpm.1994.409501
53. Investopedia (2016) Sharpe ratio. http://www.investopedia.com/terms/s/sharperatio.asp
54. Madlener R, Wenk C (2008) Efficient investment portfolios for the Swiss electricity supply sector. SSRN Electron J. https://doi.org/10.2139/ssrn.1620417
55. Liu Z (2012) Energy portfolio management with entry decisions over an infinite horizon. Appl Math 03:760–764. https://doi.org/10.4236/am.2012.37113
56. Fuss S, Szolgayová J, Khabarov N, Obersteiner M (2012) Renewables and climate change mitigation: irreversible energy investment under uncertainty and portfolio effects. Energy Policy 40:59–68. https://doi.org/10.1016/j.enpol.2010.06.061
57. EIA-DOE, EIA (2012) Assumptions to the annual energy outlook 2012
58. Schwartz ES (2004) Patents and R&D as real options. Econ Notes 33:23–54. https://doi.org/10.1111/j.0391-5026.2004.00124.x

Chapter 6
Large Transportation Project: Lessons Learned from Italian Case Studies

Roberto Maja

Abstract This chapter is aimed to the analysis of typical issues occurring during the realization of megaprojects related to infrastructures and transportation services in Italy. As it possible to expect, the variety of transportation megaprojects is extremely broad. In the first section of this paper, the author will consider some of the most significant megaprojects in the area. During the second section of the analysis, the author will better describe the features of a successful megaproject, even if it did not result with the expected outcomes: the tramway in the Bergamo valleys. After a first not exciting period of existence, this megaproject can be defined as satisfying. However, also in this case, the preliminary technical and economic analyses turned out to be excessively optimistic. This project is interesting also because it was the object of an ex-post evaluation which allows to obtain a comparison between the effectively obtained results, especially in financial and economic terms, and the evaluations made by the sponsoring actors made during the preliminary planning phase and the request for public financing.

Keywords Megaproject · Transportation engineering · Mobility planning · Tramways · Infrastructure economics

6.1 Transportation Megaprojects in Italy

"Italy needs large and small transport infrastructures". For many years (or decades) this concept has constantly been stressed by governs, political parties, media, industry representatives, committees and various stakeholders. According to someone this may be true, to others instead Italian infrastructures are even too many.

Maybe the right question to ask is another: are the transport services in Italy adapted to the satisfaction for transport demand of people and goods?

In the academic field, it has always been taught that the first two steps that needs to make in a planning process are the analysis of the transport demand and the analysis

R. Maja (✉)
Laboratorio Mobilità e Trasporti, Politecnico di Milano, Milan, Italy
e-mail: roberto.maja@polimi.it

© The Author(s), under exclusive license to Springer Nature Switzerland AG 2020   71
E. Favari and F. Cantoni (eds.), *Megaproject Management*,
PoliMI SpringerBriefs,
https://doi.org/10.1007/978-3-030-39354-0_6

of the inefficiencies of the transport network in terms of traffic congestions, shortage of services, air quality problems and energy resources consumption.

Once the demand has been estimated, alternatives projects should be made, evaluating also their costs and benefits, their environmental consequences and their social and economic impacts.

Then political institutions have to choice the projects that are able to best satisfy strategic objectives that the same political institutions have set out at different territorial levels.

A transport system consists of an infrastructure, some vehicles and a series of organizational assets aimed at the service production. So, infrastructures are only a part of the system and they shouldn't be considered an end in themselves. Of course, they attract attention, because they represent the most expensive and financially relevant part of the transport system, but the construction of infrastructures is only the very first expenditure to be incurred. Thinking to the expenditures needed for the provision of the service throughout his entire life cycle, that may last many decades, we find out that they are higher than the building costs of the infrastructure [11].

Obviously, it is necessary to consider not only the costs incurred but also the benefit obtained: a transport service is in fact essential to the survival of a human community.

Despite everything, the infrastructures are always the mayor subject of the general interest, probably because of the multiple financial resources allocation that is needed to their building.

Moreover, in many cases is impossible not to presume that the real aim of the central and local institutions is to obtain the political managing of a big amount of moneys, justifying it with the need, maybe pretexted, of building transport infrastructures.

This element can obviously generate considerable distortions: some territories are lacking transport infrastructures while in other the service is underused [11]. There are several example that can be listed: the high-speed rail networks, the "hub&spoke" airports sometimes totally underused, the regional airports often useless, motorways with any traffic, and not at least, collective local transport networks whose effective utility are often put in doubt.

The main topic of this chapter is to analyze a methodological problem: when deciding to build a transport infrastructure (or, rather, a system aimed at offering a transport service) is it usual to make feasibility study of technical, economic and social nature? The answer at this question is usually positive. However, another issue can be pointed out: some of the elements considered in the analysis are, for their nature, difficult to define in an objective way, because they are characterized by a high level of uncertainty and they can be easily, accidentally or intentionally, distorted. Some of the typical example are the overestimation of the transport demand and of the indirect benefit (of environmental and social nature) associated, and the underestimation of the costs of implementation and functioning. Usually, the final costs of realization are remarkably higher than the budget estimated in the initial project, the costs of functioning are not taken into consideration, and the demand actually satisfied is lower than expected, at least in the first period of functioning.

The overestimation of the demand is often caused by two elements: an overly optimistic assumption of the so called "modal choice", that is the attitude of a service to subtract users from the individual motorized transport; the scant attention paid to the realization of complementary infrastructure and services, i.e. the interchange car park, the pedestrian path, the location for transports stops, the information systems and the integrated pricing.

In the reality is common to find examples of really expensive infrastructures that do not generate public utility, such as the so called "white elephant" buildings.

However, it isn't rare to find virtuous examples of transport network really well designed and integrated in the territory that are able to offer an excellent service and that suffer from the opposite problem that is they are constantly overcrowded. Another typical case is the realization of transport systems that are at the beginning underused due to the inadequate complementary intervention that have been made and that, some years after their realization, become a key in the fulfilment of the mobility in the territory.

Two interesting examples are the motorway BreBeMi and the tramway of Valli di Bergamo. The first one is an infrastructure that connects the bypass road network of the two cities of Milan and Brescia [9]. This infrastructure (that is 60 km long) was realized in project financing, relying on the revenues derived from a minimum traffic of 60.000 vehicles per day. During the first years of operation (the opening was in 2014) the traffic volume was disappointing and well below the forecasted: almost 20.000 vehicles per day. This result was mostly due to the lack of the final stretch of the link to the A4 motorway near Brescia, that didn't facilitate the access to the new infrastructure. After the link construction in 2017, the traffic has significantly increased, also because several new industrial settlements and tertiary services were realized in the same area. The facts about the Valli di Bergamo tramway will be reported in the next part of this chapter.

Another typical problem that Italian megaprojects have to face is the decisional process. There are some steps of the process that are clearly unavoidable like for example the evaluation of the environmental compatibility; but the overlapping of political and administrative skills at different decision-making levels slows down the time of realization with the result that some infrastructure are already outdated the day of their construction.

The realization of the railway connection between Florence and Rome (the so called Direttissima, 250 km long), started in 1969 and was terminated in 1992. At the time of his planning it was the most modern and speed railway system of Europe (250 km/h) but when it was terminated, France had already realized, ten years earlier, hundreds of kilometers of the TGV railway at 300 km/h.

The building of the underground rail link in Milan started in 1984, simultaneously to the realization of the rail link in Zurich; this one was opened in 1989 while the completion of the one in Milan was several years later, in 2008. There's no need of talking much about the costs borne by the general public that were largely wasted on corruptive action of various nature. The service public utility of the infrastructure was indisputable but the wasteful expenditure of money cries for vengeance.

The excessive interest that is usually devoted to the realization of the transport megaprojects or, better, to the financial resources that the State spends, is balanced by the scarce attention paid to the management of the project. This, of course, is a generalized and merciless judgment, but it is suffices to mention the collapse of the Ponte Morandi of Genoa and other bridges located in Italy to understand that it is a reality also characterized by catastrophic events.

In the end, some considerations on the construction costs should be pointed out. The case of the Italian high-speed rail network is emblematic: it is characterized by a double or triple kilometer cost compared to similar networks in other European nations.

However, it should be considered that their route is parallel, for some hundreds of kilometers, to some sections of the motorway network and that for the construction of the railway it was necessary to reconstruct hundreds of overpasses and dozens of motorway junctions.

Moreover, the municipal administrations involved in the railway route intervened with their power of veto. In order to obtain their approval it was necessary "reciprocate" them with an endless series of compensatory works that had nothing to deal with the rail service. These compensatory works have been the matter of exhausting negotiations that have prolonged the realization times and have increased the overall costs. The high-speed Italian railway is still a matter of discussion. There are conflicting opinions regarding its current usefulness in relation to the costs of realization. Currently the frequency of the service is comparable to that of a suburban subway and the trains (managed by two competing railway companies) are constantly crowded, but these benefits have been obtained at great price.

However, it seems more correct to consider the costs strictly related to railway works which, although very large, are significantly lower than the total ones, which include the compensatory works over mentioned.

These additional costs shouldn't be charged to the railway technique but to the inefficiency of the administrative, managerial and political processes typical of Italy. For an analysis of the Italian and Spanish high-speed rail networks, it may be interesting to see the contribution [4].

To conclude these introductory notes, it is necessary to underline a constantly forgotten element: the ex-post analysis. Assuming that each project must have a cost-benefit assessment, it is noticeable, at least in Italy, a lack of analysis and checks carried out after the new transport systems have started to operate.

From a methodological point of view, it is considered essential to carry out new cost-benefit analysis after the realization of the projects using real traffic data and evaluating the effective benefits generated for the community.

In the following part of this chapter will be reported some considerations regarding the design, the realization and the impacts of the Valli di Bergamo tramway. Compared to its size it is not a properly megaproject, however it has been decided to consider it because it is one of the very few cases of an ex-post analysis that allow to make many interesting evaluations. The elements used for this discussion are taken from the contributions [2] and [3], to which reference is made for further details.

6.2   Bergamo Valleys Tramway System

The project examined here concerns a tramway built in 2009 on the route of a railway built at the end of the 1800s and suppressed in the 1960s.

It is a typical example of transport policy adopted in Italy and in many other European countries, that followed the following phases:

- Construction of a railway connection when this constituted the only possibility of transporting people and goods between the end of the 1800s and the beginning of the 1900s;
- Gradual decline of the railway service in the two decades following the end of the second world conflict caused by two phenomena: the spread of private motorization and the opinion, not always acceptable, that the railway service was no longer able to satisfy the increasing mobility demand;
- Suppression of local railway lines and satisfaction of transport demand with road mobility only;
- Beginning and progressive increase of road congestion, unbalanced use of energy resources and worsening of air quality and land use;
- Growing sensitivity towards environmental protection, landscape and quality of life;
- Awareness of the availability of alternative modes of transport as opposed to simple individual motorized mobility, no longer sufficient to guarantee commuting and occasional journeys with travel times compatible with user expectations;
- Tendency, started in the 1990s, to reconstruct the old rail and tram links abandoned in the previous decades and to build new infrastructures.

Moreover, the urban and social transformations have reversed the trend in urbanistic decision in the 50s, 60s and 70s, relocating residential and productive settlements in rural areas previously abandoned.

This trend has not been adequately accompanied by the creation of mobility systems suited to sustaining an increasingly widespread demand and has led to the uncoordinated construction of road infrastructures characterized by constant congestion.

6.2.1   Who Was the Project Promoter? How Was the Planning Process Set up?

The project of the tramway of Valli di Bergamo has been proposed several times since the early 1990s by the Province of Bergamo and the Lombardy Region.

The decision-making process was rather long and troubled. The final approval of the Valle Seriana tramway from Bergamo to Albino dates back to 2000 with the allocation of 37 million euros discounted to 2000. This project is included in many planning programs of local institutions, for example: the Territorial Provincial

Coordination Plan of 2004, the Triennial Program of Services 2001–2003 of the Province of Bergamo, the Bergamo Territory Government Plan of 2010, the Mobility Plan of the Municipality of Bergamo of 2008.

These plans include a very ambitious set of projects that concerns in particular the construction of a series of tramways in the more urbanized provincial territory and in the city of Bergamo itself. The Seriana Valley tramway is only part of the planned network. The project of the 10 km long Brembana Valley tramway is now at an advanced stage.

6.2.2   What Kind of Demand Forecast Was Made? Has a Modal Choice Analysis Method Been Used? Has an Analysis Been Carried Out on the Stated Preferences (SP Analysis)?

The demand forecast was carried out by three different analysis dating back to the years 1994, 2001 and 2007.

The first analysis dates back to 1994 [6] and concerns both the lines of the Seriana Valley, the only one realized, and of the Brembana Valley, currently being on project. The time reference of the estimate is referred to 2001. The Bergamo-Albino line (Valle Seriana) was inaugurated in 2009.

The forecast of potential demand for the Seriana Valley considers commuter traffic at rush hour to Bergamo and total daily traffic, i.e. commuting and occasional, in both directions. The analysis estimates two hypotheses, a minimum and a maximum one. According to the maximum hypothesis, the hourly traffic in the morning amounts to 16,000 passengers and the total daily traffic to 40,000 passengers. The first consists mainly of commuters transferred by the public service on pre-existing buses and the second consists largely of individual motorized transport users who opt for the new service. The total traffic in the minimum hypothesis is 25,000 passengers.

The second analysis dates back to 2001 [7]. The estimate is referred to 2005 and leads to a result lower than the one of 1994, equal to 19,000 users in the minimum hypothesis and 30,000 in the maximum hypothesis. The results obtained from the third analysis [8] are slightly lower.

The modal split model states the transfer of 40% of users from individual motorized transport to the use of the new tram service. Any investigations have been carried out on the Stated preferences (SP investigations) in order to calibrate the model (for a description of the theory of SP surveys see [12].

The cost-benefit analysis doesn't consider an increase in demand over the time. This hypothesis is precautionary. However, all the forecasts are very overestimated with respect to the actual traffic recorded after the tramway was built.

6.2.3 How Were Economic and Financial Analysis Structured?

In the early stages of the planning process, the Valli di Bergamo Tramway project was not subjected to an economic analysis. The first cost-benefit evaluation dates back to 2000, when the municipal council of Bergamo discussed and approved the extension of the line from Torre Boldone to Albino [10]. The official analysis was carried out by the Province of Bergamo following a classic cost-benefit procedure. The structure of the evaluation complies with the typical procedure adopted in Italy at that time and it is very simplified. The analysis does not consider some indirect non-monetary benefits, for example the environmental benefits, and is affected by some errors in the evaluation.

The direct monetary benefits have been estimated by referring to the analysis of the transport demand made in 1994 and are overestimated.

The evaluation is characterized by the following elements: taxes and transfers are excluded from financial costs; the coefficients refer to the literature of the time and are clearly reported; the discount rate used is 4% and, being rather low, it is favorable to the realization of the project; the value of time for the users of the public transport amounts to 5 euros per hour, and that of the users of the individual transport amounts to 6 euros per hour. The values are slightly higher than the ones used in that period in similar studies. The result of the evaluation is positive, and favorable to the realization of the infrastructure. The community surplus amounts to around 40 million euro discounted back to 2000. It is important to note that only one alternative project was evaluated by comparing it with the pre-existing condition. Therefore, the effects of any variants have not been analyzed, for example different route hypotheses.

6.2.4 What Are the Benefits and the Indirect Costs Evaluated in the Analysis? What Method Is Used for Their Evaluation?

The most relevant direct benefits are the reduction in travel times achieved by the users of public transport who would pass from the bus lines to the tramway, by those that would pass from individual motorized transport to the public one and by those that would continue to use their own cars.

Some indirect benefits have not been considered, in fact the analysis ignores the reduction of polluting emissions caused by road traffic and the reduction of accidents. This lack is unfavorable to the realization of the project.

The benefits deriving from the savings in transport costs by users have been calculated, but with a wrong procedure.

6.2.5 Does the Transport System Implemented Match the One Planned? What Is the Quality of the Final Service? Have Complementary Infrastructures Been Realized?

The realization of the Valle Seriana tramway is stick to the plan discussed and approved over the years and to its design.

This new transport system represents one of the most successful infrastructural works at national level in the last two decades and can be considered a good practice.

The availability of the railway line of the old Bergamo-Clusone railway line suppressed in 1967 has allowed to avoid the construction of new expensive and invasive works and to reduce construction times. The realization of complementary infrastructures has facilitated the integration with the territory and with its mobility system, for example the interchange parking lots and the interlocking of the traffic lights system.

The implementation of the service also represents a good example of coordination. The tramway timetable is scheduled to comply with connections with other public transport services. Interchange parking facilitates integration with individual transport. The pricing of the various services offered in the territory is integrated.

6.2.6 Have the Traffic Forecasts Turned Out to Be Realistic? Has the Traffic Been Overestimated or Underestimated? What Elements Led Passengers to Use or not to Use the New Infrastructure?

After the opening in May 2009, in the first few months of the year the demand served remained at rather low levels. During 2010 the annual demand reached the value of 3 million passengers per year up to the current value (2018) of 3.7 million [1], which corresponds to an average of 10,300 passengers per day. The generated traffic amounts to 535,000 tram × km × year, over a line length of 13 km [13].

The current daily demand, in working days only, stands at 13,000 passengers per day, 18% more than the demand recorded in 2010. The demand during the feast days is obviously lower, however in the winter period people going to Bergamo for leisure and shopping on Saturdays and Sundays reaches remarkable peaks, enough to cause crowding problems.

In 2010 two remarkable facts happened. In May, a major national event brought 500,000 people to Bergamo, generating over 60,000 passengers per day. This leads to two considerations: the tram system is able to withstand, in particular circumstances, considerable demand peaks. However, it's clear that the infrastructure is largely oversized, despite the constantly congested viability of the territory.

In December 2010, a heavy snowfall caused interruptions in the road system. This event allowed many users of individual transport to "discover" the availability of the tram service and this fact determined an increase in daily demand, also demonstrated by a significant increase in the number of cars in the interchange parking lots, which had previously been almost empty. However, this increase does not exceed 600 passengers per day, and shows also that it is not easy to change the habits of the population of this territory, despite the good quality of the service offered.

Overall, it is possible to draw this conclusion: the forecast made in 1994 in the maximum hypothesis was completely incorrect, amounting to 40,000 passengers per day. Even the minimum hypothesis, 25,000 passengers per day, has been blatantly disproved by the current situation. The 2007 minimum hypothesis was more realistic, 16,700 passengers per day, but it is excessive too.

This is a typical aspect of the demand analysis that are carried out when a megaproject is proposed: the forecasts are generally too optimistic and are affected by evident levels of uncertainty that inevitably leads to distort the results of the cost-benefit and financial analysis. The result is an increase in management costs that inevitably affect the entire community.

6.2.7   Making an "with Hindsight" Assessment, Have Some Mistake in the Evaluation Been Made? What Lessons Can Be Learned from Errors?

Given his financial dimensions, the Valli di Bergamo tramway, cannot properly be considered a megaproject. However, it was decided to describe it and use it as an example because it still represents a relevant infrastructure for its territory. Furthermore, it has been one of the very first realizations of new tramways in Italy for at least six decades.

Finally, a study carried out in 2012 at the Politecnico of Milan [2, 5] is available, which compares the cost-benefit analysis carried out at the time of the presentation of the project with the results of an ex-post analysis carried out two years after the entry into service. This is one of the very rare examples, in Italy, of comparing forecasts and actual operations in financial and economic terms.

The Traspol Laboratory of the Politecnico of Milan has reconstructed the original evaluation model of 2000, correcting some methodological errors and then reproducing it using the actual traffic data recorded in 2009 and 2010.

The first ex-post analysis referring to 2010 contains some corrections with respect to the ex-ante analysis of 2000: actual investment costs which are higher than those planned; the service offered in terms of trams × km; the number of users gained from the existing bus service and from individual motorized transport (modal split); the savings deriving from the elimination of the bus lines replaced by the new tramway in terms of bus km and euro per km; the time saved by users compared to the original distances; the value of time reformulated in more realistic terms.

The data that has had the most significant change concerns the demand served, which in the 2000 analysis was expected to be 25,000–40,000 passengers per day, but in 2010 still did not exceed 9,000 passengers per day.

Furthermore, the 2000 analysis also contained some assessments that were not favorable to the realization of the tramway. The unit benefits were estimated at around 1000 euros/passenger, while in the 2010 reconstruction they were revalued to almost 3000 euros/passenger.

The new analysis produced the following results.

The original NPV of 2000 was around 40 million euros; the NPV recalculated with the demand and the real costs is worth −110 million euros, that is a significantly negative value that would not have allowed the realization of the infrastructure!

The NPV calculated with the 2010 ex-post model, containing a revaluation of direct and indirect benefits, would have led to a much higher NPV of around 190 million euros. This data is lowered to −48 million euros if the original forecasts of demand and costs are replaced with the real data. It is therefore a more favorable value, but still a negative one.

A further recalculation of the NPV carried out with the expected demand data for 2011 led to an even more favorable, but still negative, assessment of −25 million euros.

In summary, the original analysis of 2000 contained two significant errors: the underestimation of indirect benefits related to the reduction of polluting emissions and the considerable overestimation of the demand that could be obtained from the tramway and from the modal split. The final result is however adverse to the feasibility of the project.

However, Traspol also carried out another analysis, researching the values of the demand that could possibly overturn the estimate of the NPV and formulating the following hypotheses: the acquisition by the tramway of a significant part of the motorized passengers coming from the upper valley upstream of the Albino terminal could increase the benefits in terms of time saved; served demand value that exceeds 9200 passengers per day, of which at least 25% comes from the modal split could increase financial revenues and consequently the NPV. Examining the 2018 financial statement [13] it seems that one of these objectives has been achieved, in fact the current demand is around 10,300 passengers per day. However, no data is available about the modal split and the time saved.

6.2.8   Final Considerations

To sum up, the following conclusions can be drawn from the ex-post analysis of the Valli di Bergamo tramway project, which can be considered instructive to evaluate the typical factors of many other transport systems and services projects.

The overestimation of the demand led to an opinion that was too favorable to the realization of the project, which was disproved by reality. This is unfortunately a common vice of many similar projects and in particular of many megaprojects.

For what concerning the forecast of the potential demand it doesn't seem that the calibration of a modal choice model has been made using the technique of surveying the Stated Preferences, which is suitable to provide indications on the possible behavior of the potential users with respect to the features of the service offered. A modal choice model should provide in particular the users' "willingness to pay" (the demand curve). Through this factor it is possible to evaluate, for example, the share of possible users who would be willing to pay the pass ticket for a new service in order to have advantages in terms of time and cost savings compared to the use of its car.

In the analysis of the benefits some have been omitted that would have increased the NPV, although leaving it negative. This fact can lead to two considerations: the error was contrary to the realization of the project, however the tramway was realized anyway because these missed benefits were bi-launched with the supposed benefits expected from an overestimated demand; the economic evaluations, although necessary, often prove to be susceptible to methodological errors or interpretative discretions that raise many doubts about their reliability and about the attention with which they are read and considered by policy makers. Many megaprojects (for example, see the Italian case of the so-called TAV Turin-Lyon) have been subjected to numerous technical and economic evaluations made by different analysts, who have provided conflicting results, some clearly favorable, others drastically opposed to their realization.

References

1. Agenzia per il trasporto pubblico locale del Bacino di Bergamo (2018) Programma di Bacino 2018. Bergamo
2. Beria P, Borlini A (2011) Il nuovo tram di Bergamo—Valutazione ex-post delle stime di domanda e dell'analisi costi benefici. In: Traspol Working Paper. www.traspol.polimi.it, Milano
3. Beria P, Borlini A, Maja R (2012) Il nuovo tram di Bergamo—Una valutazione ex-post. Territorio, Milano
4. Beria P, Grimaldi R, Albalate D, Belc G (2018) Delusions of success: costs and demand of high-speed rail in Italy and Spain. Transp Policy 68:63–79
5. Borlini A (2011) Metrotramvia Val Seriana—Nuove prospettive trasporto collettivo. Master thesis, Politecnico di Milano, Milano
6. Centro Studi Traffico (1994) Sistema di Trasporto a guida vincolata per i collegamenti fra Bergamo e le Valle Seriana e Brembana—Valutazione della Domanda Potenziale. Allegato A, Provincia di Bergamo
7. Centro Studi Traffico (2001) Tramvia delle Valli Bergamasche—Tratta Bergamo-Albino—Definizione della domanda potenziale. TEB Bergamo
8. Centro Studi Traffico (2007) Studio della domanda e sugli effetti indotti per lo sviluppo della rete tramviaria-ferroviaria. TEB, Bergamo
9. Garzarella A, Perotti M (2019) Benefici diretti e indiretti dell'autostrada A35 Brebemi, Agici
10. Marcotti G (2000) Tramvia delle Valli. Analisi costi-benefici e piano economico finanziario, Provincia di Bergamo

11. Ministero delle Infrastrutture e dei Trasporti (2019) Ufficio di Statistica, Conto nazionale delle infrastrutture e dei trasporti. Roma
12. Ortuzar JdD, Willumsen LG (2015) Modelling transport. 4th edn. Wiley
13. Tramvie Elettriche Bergamasche SpA (2018) Bilancio d'esercizio 2018. Bergamo

Chapter 7
Milano Expo 2015 Regeneration Process: Between Legacy and Megaproject

Sara Protasoni and Michele Roda

Abstract Essay main goal is a critical discussion about prospects and methods of the development—intended simultaneously as regeneration, redevelopment and reuse—of Expo 2015 area. The asset seems to be an exemplary case-study of megaproject, where at least 3 interacting aspects could be highlighted, all of them capable of defining a specific focus, along the complex ridges among architecture, planning and landscape: the emerging need to design a neighborhood/city *ex novo*, in a very brief time frame; an overlapping of highly attractive public and private functions, with a strong international vocation; a close integration between urban design and landscape project, as key factors of the quality. The analysis proposes a method based on a complex evaluation, structured around some nodal elements applied: time, urban structure and tissue, design of open spaces, integration, actors involved. The result will allow to define a clear method of reading and evaluating complex projects, characterized by the participation of various stakeholders, not separating it from the landscape, urban and architectural project elements and components. The cultural background is characterized by the strong conviction that they need to be supporting factors, and not secondary ones, in the processes of decision-making about territorial transformations.

Keywords Regeneration · Public/private · Expo 2015 · Landscape · Design

7.1 Preface—Introduction

Essay main goal is a critical discussion about prospects and methods of the development—intended simultaneously as regeneration, redevelopment and reuse—of Expo 2015 area, in Milano northern periphery.

S. Protasoni
Department ABC, Politecnico di Milano, Milan, Italy
e-mail: sara.protasoni@polimi.it

M. Roda (✉)
Department DASTU, Politecnico di Milano, Milan, Italy
e-mail: michele.roda@polimi.it

© The Author(s), under exclusive license to Springer Nature Switzerland AG 2020   83
E. Favari and F. Cantoni (eds.), *Megaproject Management*,
PoliMI SpringerBriefs,
https://doi.org/10.1007/978-3-030-39354-0_7

The asset—considering its own history (although recent), its dimensions and its implications—seems to be an exemplary case-study of megaproject (1.114 milions of €, in the period 2009–2015, of public investments + about 1 milion of € invested by guest countries in the private pavilions), where at least 3 interacting aspects could be highlighted, all of them capable of defining a specific focus, along the complex ridges among architecture, planning and landscape:

- the emerging need to design a neighborhood/city *ex novo*, in a very brief time frame, with the constraint/opportunity to define a new role and a new meaning for the exhibition site fragments, in relation to the original context;
- an overlapping of highly attractive public and private functions, with a strong international vocation;
- a close integration between urban design and landscape project, as key factors of the quality, mainly regarding the infrastructures (gray, green and blue) role and their functional, ecological and landscape value.

A deep discussion about the experience of Expo Milano 2015 and its legacy becomes more and more topical looking at the decision, announced in June by the International Olympic Committee (IOC), to give Italy the responsibility to host the Winter Olympic Games of 2026. These are the words supporting the candidature of Milano-Cortina: *The Milan-Cortina 2026 Candidature fully embraces the IOC's Olympic Agenda 2020/New Norm, putting the long-term development strategies and challenges of the host cities and regions at the centre. We will contribute to a golden decade of Olympic and Paralympic sport, placing Milan-Cortina alongside other leading international cities, to help reposition the Games in modern society. The selection of the venues is fully aligned with the ambition of northern Italy to become a world-class hub of excellence for winter sports. [...] The Games will also be conducive to greater cooperation in the entire Alpine macro-region, to improve the attractiveness of the mountains as a place to live and reverse the trend of depopulation. Finally, the Games are a unique opportunity to showcase Italy's design style that has seen the "Made in Italy" brand become synonymous with excellence all over the world.* (olympic.org/milano-cortina-2026)

It's a multi-city (or multi-region) candidature, an innovative format in such type of Great Sport Event. Milano—city we're discussing in this essay—will host:

Skating and short track games—Assago Forum (existing)

Ice hockey matches—Pala Italia (new, in Santa Giulia area, for 15.000 people) + Palasharp (to be renovated)

Main Olympic Village—Porta Roman (new)—to be reconverted after the Games into a student housing district

Medal Plaza—Piazza Duomo

The approach of our country—intended as political and social system as well as economic assets—was very controversial during the last 30 years and involves deeply also urban and architectural design field (Table 7.1).

These events were characterized by complex processes with diverse physical and spatial results [1]. Some of them entered also in the history of architecture and

Table 7.1 Great eventes in Italy

Year	Event	City
1990	Football World Cup	different venues
1992	International Exhibition Colombo '92	Genova
1994	G7 Summit	Napoli
2001	G8 Summit	Genova
2004	European Capital of Culture	Genova
2006	Winter Olympic Games	Torino
2009	Swimming World Championships	Roma
2015	Expo	Milano
2019	European Capital of Culture	Matera

planning for positive and impacting outcomes in the host city. It's possible to quote in this sense, the strong regeneration (a sort of a new Rinascimento) for Napoli, approaching 1994 Summit of G7 or, just a few years earlier, the analogous process of Genova with the rediscovering of its own sea-front (designed by Renzo Piano). Both cities lived in the Nineties a lucky period, able to transform their image itself.

But, on the other side of the coin, it's possible to remember the 2009 Swimming World Championships in Roma with the un-finished International Aquatic Centre in Tor Vergata designed by Santiago Calatrava, iconic emblem of a wrong process, not completed with a strong unbalance among money invested and results.

Also turning the look to the international level, it's possible to see how recent Great Events were characterized by very controversial results (Table 7.2).

In some cases the Great Events have been fantastic and unique opportunities to attract investments and energies oriented at urban large scale transformations [2]. Among the most well-known experiences in this line it's the Barcelona case. Without the 1992 Olympic Games, the Catalan Capital couldn't be the city we can see today: a vibrant European capital, with architectural and urban design projects able to impact on the history of architecture itself. In some other cases instead, the result is very problematic. For Greece, a contributing factor to the country's 2011 debt default was the 2004 Olympic Games in Athens. Besides some significant and positive urban transformations, the legacy of the cost overruns and incurred debt was

Table 7.2 International recent great events strongly impacting on cities

Year	Event	City
2004	Olympic Games	Athens (Greece)
2008	Expo	Saragoza (Spain)
2010	Expo	Shanghai (China)
2012	Olympic Games	London (UK)
2014	Winter Olympic Games	Sochi (Russia)

a strong weakening of the national economy in the years before the 2008 international financial crisis.

In many situations, the strongest critiscism emerged after the Event's closing. This means that it's exactly the post-event phase the most dangerous and risky one. It's the period when the lack of a sustainable and progressive horizon—maybe hidden by the enthusiasm provocked by the event itself—can turn into a great failure. With an area, urbanized and strongly transformed, rapidly becoming a no-man's-land.

So, discussing about mega-projects, we need to look steps and passages to understand the role of urban and architectural quality within such large scale transformations. The aim of this essay is deeply related with the stressed points in order to highlight and enhance consciousness about the potentials of architecture and urban design addressing transformations [3]. And in this line what happened and what will happen around Milano Expo Area seems a very topical case-study [4].

The interested area (1,2 milion square meters large) is in the northern sector of Milano metropolitan area, spread in 3 different municipalities: Milano itself, Rho, Pero. It's surrounded and defined by a crown of infrastructural lines: Milano-Torino railway in the south, A8 motorway in the northern edge. Another motorway, A52, is defining the west border, which is the one dividing the area from the Fiera district fairgrounds. Up to 2010 the area is mainly an open sector, with agricultural active fields. Milano is appointed as city hosting the 2015 Expo in 2008. After a long discussion, very difficult and politically controversial, the institutions chose this area (which was private) because of its identity and closeness to infrastructures for the temporary Expo (Table 7.3).

By a landscape, urban and architectural point of view, the area is intended as a citadel. Despite the horizontal layout (with the main axis on west-east direction), urban choice was to build a Roman castrum system with a *cardo* and a *decumano* impacting on the pavilions layout and distribution. Around the complex and not linear borders and edges, a water canal was intended as a landscape factor able to create diversity (Figs. 7.1 and 7.2).

About 4 years after the last closing of Expo site, the area is object of a strong process of transformations, coherent with the original program of re-using the site as a new hub for culture, technology and innovation, in a strong synergy among public and private investments (mindmilano.it). Up to now just 2 building-sites are open (Galeazzi Hospital and former Palazzo Italia within the Human Technopole), there's a light regeneration process in progress with some buildings (the great part is already dismantled). But the strongest phase of transforming will start from 2026 onwards (lendlease.com) (Figs. 7.3 and 7.4).

Despite the very brief period, the history and the dynamic of Expo area evolution is very complex and it strongly impacts on the actual conditions [6] (Table 7.4).

Table 7.3 Main datas of Expo experience [5]

Opening Period	May 1st—October 31st	2015
Visitors	22	milions
Investments	2.254,7	milions €

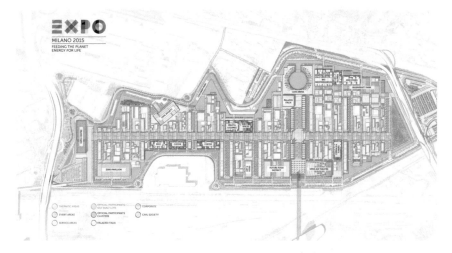

Fig. 7.1 Expo Milano 2015 masterplan

Fig. 7.2 Expo Milano 2015 crowd entering the exhibition

About halfaway through the Expo closing and the activation of the concession to Lendlease, we face a complex process of delayed uses of the spaces with some building sites already active:

- for the new Galeazzi hospital center; for the transformation of the Italian Pavilion (along the Cardo) into an Innovation Center. It's open the Design Competition for the headquarter of the Human Technopole;

Fig. 7.3 Actual images of Decumano, the main 1.5 km long axis designed to let Expo visitors to walk and to move along the sequence of Expo 2015 pavilions of Expo

- for the temporary and partial re-use of some pavilions (mainly services, not the national ones) in the pre-2026 phase. Within some months also building site for Scientific Schools of Statale University will be opened (Fig. 7.5).

The Milano Expo site experience is here presented as a topical case study of megaproject, partially in progress and building, because it shares some points and characters with megaprojects, as presented and discussed by Flyvbjerg [7]. These are the stressed points:

1. due long planning horizons and complex interfaces, it is inherently risky;
2. the project and the process, also because of its own strong political and media prominence, was led by planners and managers without a deep domain experience, in some moments leaving a weak leadership;
3. decision making, planning and management has been multi-actor processes involving multiple stakeholders, both public and private, with conflicting interests;
4. technology and designs have been (and are) non-standard. This impedes a correct dissemination of the results;
5. frequently there has been an overcommitment to a certain project concept at an early stage, leaving analyses of alternatives weak or absent and leading to escalated commitment in later stages;

Fig. 7.4 The construction site of the Galeazzi Hospital Center which foresees the activation, from 2021, of the hospital activities in the 150 thousand square meters under construction. The tallest buildings will reach 85 meters, the investment is over 200 million euros

6. due to the large sums of money involved, principal agent problems and rent-seeking behavior are common;
7. the project scope or ambition level have changed significantly over time;
8. we can see misinformation about costs, schedules, benefits and risks throughout project development and the decision-making process. The result is cost over-runs, delays and benefit short-falls that undermine project viability during project implementation and operations.

7.2   Emerging Topics

Trying to discuss the case-study focusing landscape and architectural implications and impacts, 4 nodal elements are emerging within the very complex condition:

1. time, intended as a key factor in the progressive overlapping of functional destinations and assets;
2. the urban structure and tissue with its strong and recognizable shape;

Table 7.4 Timetable of transformations and main steps

Period	What happens
31st March, 2008	The Bureau International des Expositions (BIE) general assembly in Paris decided in favour of Milano candidature
2008	Expo 2015 Society was founded and, after a long political confrontation, Lucio Stanca was chosen as its commissioner (he will be substitued later by Giuseppe Sala, who will direct the society during to the exhibition phase)
October, 2008	The Program Agreement for the definition of the urban planning discipline, with some defined points also about its legacy, is signed
2008	The Board for Architecture for Expo 2015 is named. It's composed by Stefano Boeri, Richard Burdett, Joan Busquets, Jacques Herzog and William McDonough. The goal is to produce a Director Document defining rules for drawing up the Program Agreement, a call for tenders for a Design Competition and a plan for reusing the areas. They're objectives not effectively realized
2009	It's the year of the strongest political discussion about the Expo 2015 society typology
September, 2009	The first Expo 2015's concept was presented. It was designed by a committee of four architects: Stefano Boeri, Richard Burdett, Mark Rylander and Jacques Herzog. The main idea concerns the building of a planetary garden, which sounded as an innovative layout for an Expo. The concept—different from the one presented in the candidature—will be not realized
1st May, 2010	Expo Milano "Feeding the Planet" is officially registered at BIE. It happens without the availability of the areas (they're still private)
September, 2010	The urban program with the rules for the "Expo period" and for the "Post-Expo period" is approved. The masterplan and the characters of the site are quite different from the ones presented to BIE for the candidature and also different from the first concept: the Expo model is more traditional and less innovative
16th April, 2011	The Program Agreement Conference shares the regional proposal to set up a dedicated company in order to acquire ownership of the areas of the exhibition site (300 milions of €) and to develop its post-Event use: it is the Arexpo S.p.A.
2011	due to the tight timeframe to start the activities, the permanent "Services Conference for the approval of the projects of the 2015 Universal Exposition site" is established. A procedural modality is approved with integrated contracts to guarantee speed of entrustment and realization. It is a choice, inevitable considering the delays, which also affects the procedures for the awarding of designs and works, characterized, even at the risk of quality, by a scarce use of competitions and open procedures
2013	Arexpo publishes a call for tenders for interest in the development of the area. The proposal is a sports citadel, to be implemented in partnership with Milan and Inter, the 2 professional football teams of the city. The hypothesis does not take off, leaving open the discussion on the use of the post-Expo area. An indecision that will last well beyond the closing of the Expo (arexpo.it)
2017	Lendlease has been awarded the first stage Consultancy Agreement for the Public Private Partnership which will develop the site. It's the first and necessary step for the Expo-site whole redesign (lendlease.com)

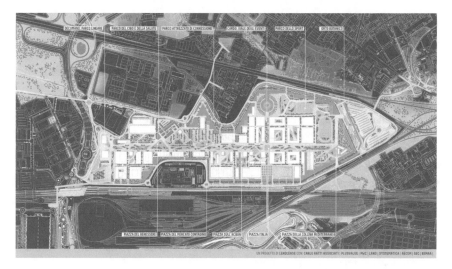

Fig. 7.5 Lendlease is the concessionaire society, from 2026, of the Expo area. It will become, in the real-estate company vision, the new Italian headquarters for international leading societies and agencies, mainly operating in the fields of technological innovation and research. In the masterplan recently approved there is a strong integration between public and private activities

3. design of open spaces as an essential element capable of bringing quality throughout the system;
4. integration/contamination of functional assets (facilities for living, production activities, services, gray, green and blue infrastructures…) and stakeholders/actors/interests;

Next pages will describe these 4 components of our discussion and evaluation.

7.2.1   Time

The temporal factor significantly impacts on designing architecture and making cities. Transformations have a long time, buildings can modify their functional programs by transforming their forms and surviving through eras, political and economic regimes. Today—and Italy is a tragically glaring example in this sense—the long times of projects, between complex processes and often inextricable bureaucracies, often lead to very contradictory results with drawings and proposals that are already old and not coherent before being built. Sometimes they're even no longer able to interpret the sense of innovation that links the project itself. Along this line of interpretation, the experience of Milano Expo, like most of the Great Events, marks an opposite moment, almost paradoxical [8]. This sounds as a reverse process:

- the development plan concerns an area that, with the exception of the centrality of some functions on the margins (such as the Fair) and of the very high level of accessibility (along the infrastructural axes, both as railways and roads), could not boast any kind of specific identity. It was a marginal and peripheral area, untouched by development projects;
- the area becomes part of the political agenda with a typically top-down process: its purchase (through a company with public participation) is discussed—not without strong controversy and tension—only within political institutions and boards, without forms of participation;
- this process happens with a very strong pressure, including media;
- the design phase, and the approval of the documents for the construction, lasts a very brief period, above all if we compare it to the normal times of far less complex processes. This contraction has consequences on the insufficient comparison among possible alternatives, in a sort of centralized dirigisme. Moreover the phases of consultations and attempts (for the most part aborted) of forms of participation result in a stalemate;
- equally short is the construction phase—which was object of a very strong social and media attention—which radically transforms the structure of the area in the short term of a few years. A real city—though designed to be temporary—is built in an incredibly short period of time;
- being the emblem of globality makes this temporary citadel (intended to be open just for 6 months) one of the places of Milano with the strongest identity. It was visited by over 20 million people, it was intensively told in the media, it became a new urban reference point;
- closing of the Expo site meant—largely respecting the original program—the demolition of a large part of the built structures. The demolition phase is significantly longer than the construction one (although in a quite rapid process) and we can face today the reduction to a sort of skeleton structure that should be, in the ambition, the armor of the "new" city;
- the current one (up to 2026) is an interregnum phase, with the area—which has a strong, and predominantly positive, identity—subject to diversified processes, both public and private: some redevelopments in progress (Mind), some building sites (Galeazzi), some projects (Statale University), some buildings and some areas occupied and used, even within the framework of temporally defined initiatives [9];
- although some important public uses will be established earlier, the time first horizon is 2026, with the acquisition of a large part of the former Expo area by Lendlease for the development of an innovation citadel, in strong synergy with the other installed functions;
- but also the medium and long-term future of the area seems to tend to emphasize the more paradoxical—and therefore interesting—elements of the area. Alongside some public functions being established (the Galeazzi, Mind, the universities), there will be many private areas, in concession to Lendlease for 99 years. This means—looking at the future, i.e. the look we need discussing about urban scale processes—that from 2125, or 110 years after Expo, the theme of transformation

and reconversion—managed by the public authorities—of the Expo 2015 area could come back [10] (Figs. 7.6 and 7.7).

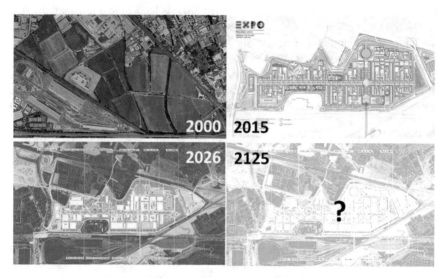

Fig. 7.6 Brief diagram of territorial transformation within the time frame

Fig. 7.7 The urban concept of the new system

7.2.2 Urbanity

Also for this second parameter it is perhaps appropriate to start the discussion from a paradoxical aspect. The temporary citadel of Expo is located in a substantially free area, but with a very articulated shape as well as being extended in a mainly east-west direction, with margins defined mainly by infrastructures, property boundaries and built-up areas excluded from the transformation processes.

This place is structured (and infrastructured) looking forward to a strong sense of urbanity, through a double axis to simulate the traditional layout of the Roman *castrum*, that is a *decumano* in east-west direction (offset from the main entrances) and a *cardo*, obviously orthogonal, opportunely placed where the area is wider.

In addition to being efficient spatial devices for the distribution of a crowd of people, which, in some days of opening, has largely exceeded the 200 thousand units, *decumano* and *cardo* serve in different ways the lots directly facing them: with long and narrow layout (almost reinterpretations of Gothic plots) along the *decumano*. In this way each guest country could substantially develop a similar program: short façade with great visibility and development towards the inside. The layout lets also a long sequence of "shared" spaces along the *cardo* which leads—also here in a logic of misalignment—to Piazza Italia with its Palace. In the areas farthest from this cross of a contemporary foundation, the common pavilions and clusters are placed as well as, the only exception in the orthogonality of the system, the pre-existing Cascina Triulza.

The margins of the *enclave* are occupied by a sort of artificial canal, like a renewed and contemporary moat—for technical and technological reasons more than with defensive intent (a need effectively guaranteed by high metal fences)—with the water intended mainly as a useful element for the climatic functioning of buildings.

The paradox mentioned above is to use a strongly urban model, as it is a foundation cross, as a model of spatial and distributive development of a city/citadel, temporary and not intended to host citizens and inhabitants but only visitors, protected and closed at night and composed by fragments largely destined for demolition after a 6-month useful life.

The strong and concentrated investment on the exhibition site is actually vastly lower, in a ratio of around 1 to 5, compared to public investments in the wide urban sector, largely of an infrastructural nature. It is precisely those 9–10 billion € of new roads and new services that today allow us to consider the area as a strong urban centrality, permitting to use the settlement model for the Lendlease project, winner of the concession of the area for 99 years starting from 2026.

There is another aspect, which the recent post-Expo have effectively portrayed and underlined: the success of the initiative (in terms of number of visitors but more generally of the national and international reception that the event received) has contributed to creating a climate of trust about the site itself, placing Milan in a position that seems unusual and unexpected: a leading city not only in northern Italy but throughout the country, a true international hub, capable of draining resources to all other cities and having a positive GDP despite the difficult situation at national level. It's a phase of optimism that on the one hand leads some theorists—few in truth—to theorize about city-state, on the other it is provoking, already in the short post-Expo period, a positive response to the planned new development of the area (wired.it).

Following the result of the competition held by Arexpo, Lendlease, in October 2015 was named to develop the area. The self-description of the society can help to focus ambitions and perspectives (lendlease.com):

Lendlease is an international property and infrastructure group with core exper-tise in shaping cities and creating strong and connected communities. Being bold and innovative characterises our approach and doing what matters defines our intent. We create award-winning urban precincts, new communities for older people and young families just starting out, retail precincts, and work places to the highest sustainability standards. We are also privileged to create essential civic and social infrastructure including state-of-the-art hospitals, universities and stadiums around the world.

The background is a *politically-correct* integration of innovation, social equity and economical development. The group—which has already developed a post-event recovery project in the case of the London 2012 Olympic Games and which in Milano has already partially operated in the recovery of the Santa Giulia district—presented itself to the race through its Italian company Lendlease Italy. In the same team there are more partners among which, for the physical and spatial sector, Land, the Milano based landscape architects firm, Sinergetica for mobility design and Carlo Ratti's studio based in Torino for the architectural project concept.

Leaving to the following points the description of some specific aspects of the project, with respect to the themes highlighted, it seems appropriate here to underline how the process of progressive replacement of the temporary functions of Expo with the definitive ones (at least in the medium-long period of the 99 years of the concession) happens—according to the preliminary masterplan presented during the tender for the area—with a settlement strategy focused on few points:

- confirmation of the urban structure based on *cardo* and *decumano*;
- confirmation of the prevailing solution (open spaces + water elements) for the marginal crowns of the area;
- replacement of elongated lots with predominantly square lots with large building blocks.

Even in the provisional nature of the proposed tender design (which is defined and refined, up to 2026, also in line with the program of operators interested in settling here), the desire to transform the area confirms its essence as the sum of different parts, as was the Expo citadel, in a sort of correspondence between spatial element and identity vocation: the *Decumano* is the new Linear Park, the *Cardo* is the Avenue of Events, around the Cascina Triulza there are the Park of Food and Health and the Piazza del Benessere, just beyond the Piazza del Mercato Contadino which forms a system of squares/public spaces together with the one on the water, in Piazza Italia and in the Collina Mediterranea. To the east there are the Sport Park and the Botanical Garden.

Even through the re-proposition of these "typologies" of spaces, the design seems to keep and preserve its strong identity of a citadel of entertainment.

The result is a very particular city, which the designers describe through 3 key words (lendlease.com):

The linear park—the masterplan finds its center in the Decumano, one of the symbols of Expo 2015. In the new district this will turn into a linear park over 1500 meters long, one of the largest in Europe, around which life will unfold city daily

Self-driving mobility—the pioneering spirit of the project extends to the field of mobility: through a gradual program of conversion of spaces, starting from the Decumano, the roads of the Science Park of Knowledge and Innovation will come to accommodate cars driving autonomous, in advance of what will happen in more and more metropolises in the near future. In this new scenario, sharing a vehicle will become more and more frequent, and it will be possible to reduce traffic, improve environmental quality, and encourage the creation of new business and work opportunities.

The Common Ground—following the principle that the health of a neighborhood depends largely on the vitality of its public spaces, the masterplan proposes to establish a Common Ground—a two-story space at street level that winds through all the project areas—on which are going to be alternated squares and pedestrian areas, orchards and gardens, shops, laboratories and court buildings, in a continuous exchange between open and closed, public or more collected environments.

Even in these descriptions, the slogan and image dimension seems to prevail. A dimension inevitably linked—with reference also to the previous point—to the process of building cities in a very short time and, substantially, with only one major operator and not a plurality of actors in the field. This obviously produces a coherence, also linguistic and of approach, deeply reducing the complexity (Fig. 7.7).

7.2.3 Open Spaces

The topic of open space—also in an often contradictory and demagogic form—has characterized the debate around Expo since the beginning of its planning. The large park (urban? agricultural? of prevailing landscape value? sport?) has been one of the told and repeated *mantras*. It should—with a lot of urban planning tools to confirm it—occupy at least half of the surfaces and become (also) a compensation factor for local communities, worried about the potential impacts related to the transformation of the area and the settlement of the Expo citadel.

More than 50% of the surface will become a park.

In the repetition of principles, even more when it concerns quantitative aspects, the sense of general architectural and urban quality is often lost. Already the Expo citadel has shown a utilitaristic and functionalist approach to open space design: a support device for the built-up area, sized on the expected flows rather than on the search for its intrinsic quality and habitability, sequence of fragmented and disjointed places, definitions (park, hill, garden) that seem destined more to give suggestions than to impact on the construction of space and landscape.

Even in the preliminary level—and hopefully subject to significant changes in the subsequent architectural fallout—the Lendlease masterplan proposal also does not seem to redeem this dimension. Some recognizable figures of the open space emerge (landsrl.com):

- the *decumano* of Expo, dismantling its own roof structures and tents, is redesigned as a large linear park with a sequence of paved parts and green islands; it builds an identity of open space that will become the backbone of the proposed light and sustainable mobility system;
- the area around Cascina Triulza is the only one where the open space prevails hierarchically over the subdivision in plots. It is an "urban" green, closely related to the accesses (the north in direction of Bollate), the only one in which the water of the canal is redeemed from a purely functionalist dimension to try to become a factor of the urban quality;
- the third sector where the open space seems to emerge as a figure is the eastern one. Here the choice seems to be identity and functional, it is in fact a sort of park equipped with sports fields and a botanical garden, in a design layout characterized by large islands (Fig. 7.5).

This synthesis has the objective to communciate how the vision of open space, obviously understood in a broad and wide sense, and emerging in the double—close-range—planning, is aimed at the construction of a support element, able to give functional answers but not to express its own autonomy and power. There does not seem to be the will to design autonomous spaces capable of being, with different hierarchies of naturalness, recognizable elements even on a large scale—including the territorial systems and their own infrastructures. We can say that a high vision, a long-range strategy, misses: it's difficult to get the will to build landscape systems. And in this weak aspect we can perhaps read a continuity with the most significant "renunciation" of the first planning and construction phase of Expo: the so-called land and water routes that should have been territorial systems linking the city and the site.

The Via *d'Acqua*, in particular, seems to represent the stronger contradictions of the process. It was the flagship of the candidacy, a new irrigation channel of 21 km in length to bring water from the Rho-Pero site to the Naviglio Grande, to the south, up to the Agricultural Park. It had to be the tangible sign that Milano thought its Expo in terms of sustainability and protection of resources, a sort of renaissance of mobility on water consistent with the path (never interrupted but never even strongly undertaken) of a renewed navigability of the canals. With the approach of the Great Event the idea is strongly resized: no longer a navigable vessel, but a spillway canal, the same length, but with different caliber and impacts, along the western edge of the city, through the crown of existing parks (delle Cave, Pertini, Trenno and Bosco in Città). An idea, ultimately, of little ambition and even less suggestive, and not built, neither in its low scale dimension. The canal that surrounds and will surround

Fig. 7.8 Aerial view of Lendlease proposal, with the green areas planned

the site seems to be the result of this whole process: neither river nor element of the collective space (Fig. 7.8).

7.2.4 Integration

The area was designed as a largely public investment, for the purchase of the areas and therefore for the construction of the surrounding infrastructural works as well as for the infrastructure of the area itself. In fact, the investments of Expo countries for the construction of self-built pavilions are largely public.

The post-Expo phase, as well as being implemented, is characterized by a marked integration between public and private capital. Public funds are necessary for the realization of the Human Technopole, which will be born along the *Cardo* road, where Palazzo Italia is also located, for which the competition phase is currently operational (it's one of the few published for the Expo area). For this structure the State invests 150 million a year for 10 years. When fully operational, in 2024, there will be 1500 employees, who will use 35.000 square meters and will be involved in advanced genome research.

The other public investments are the establishment of the scientific faculties of the State University of Milano, which will have to find between 340 and 380 million, occupying 150 thousand square meters and the new Galeazzi hospital (active site), which won a tender for 25 million.

The economic impact of the public sector, according to the estimate of European House Ambrosetti, is about € 6,9 billion.

The investment for the private part is worth about a third, estimated at around 2 billion, and managed by Lendlease with the involvement of about 50 companies that have expressed their interest in settling on the former Expo site. The project is a multi-phased, mixed-use redevelopment that is expected to include commercial, residential, retail and public realm. The work could begin as early as 2021 and is planned to take 15 years to be completed.

MIND (acronymous for Milano Innovation District) is one of a number of major urbanisation projects Lendlease currently has underway within Europe. Others include London's Elephant Park, Euston Over Station Development and Silvertown Quays. Dan Labbad, the CEO for Lendlease Europe, said: *"Work is already underway on our plan to create an exceptional new district in Milan. The Milan Innovation District is a scheme with huge potential and one where, working alongside our partner Arexpo, we are looking to build a world-leading centre of scientific excellence alongside an outstanding and active mixed-use neighbourhood"*. Andrea Ruckstuhl, Head of Italy & Continental Europe, adds: *"We are delighted to have been chosen as partner for such a visionary city project. We look forward to working in partnership with Arexpo in planning the first stage of this very significant urban renewal. The Expo in 2015 was a very positive event, bringing together millions of people from across the world to experience a showcase of technology, innovation and culture. We look forward to combining the very best local and international expertise to realise Arexpo's vision"*. By the public part, Giovanni Azzone, president of Arexpo, talking about public-private partnership ruling Expo development, declared:

This agreement is another important step forward for the Milano Innovation District, the area where the Expo Milano 2015 took place. This collaboration between Arexpo and Lendlease will provide the foundation of a top level international project, and a unique opportunity for Italy. (mindmilano.it)

The tender for the areas, won by Lendlease, quoted the building of a masterplan for the *Science, Knowledge and Innovation Park* which also included the most properly public parts, such as the Human Technopole headquarters and the Campus of scientific faculties of the University of Milano. The objective of the call was the search for a technical, economic and financial operator capable of supporting Arexpo in developing and enhancing the Expo 2015 site. The relationship among the public company and the private operator consists of a partnership through a mixed contract of contract and concession. The Lendlease proposal prevailed in the 2 phases of the tender: the first with the technical, economic and financial advisory activities in support of Arexpo, for the definition of the masterplan of transformation of the area, and for the conception and elaboration of the relative economic and financial plan. The second phase, instead, related to the implementation activities, therefore also the design and construction activities of the works, and the management of the spaces.

The contract will provide for a surface right in favor of the concessionaire for an extension of gross floor area of not lower than 250 thousand to 400 thousand square meters. The area is already infrastructured. Giovanni Azzone, president of Arexpo, explains the competition (arexpo.it):

> Our goal was to create a project in the former Expo area that has no equal in Italy and that can be compared with the most virtuous examples of urban transformation at the international level. We have chosen to focus for this place on the great themes of the future: public and private research in the field of health, well-being, care for people and to combine this project with a large university center. Carlo Ratti, on the other hand, one of the signatures of the masterplan presented by Lendlease explains (carloratti.it):

> The former site of Expo 2015 will become a place to experiment with new ways of working, of doing research, of living, of being together and moving. A garden city with which to imagine a future modeled and reshaped on the needs of its inhabitants. This area offers us an extraordinary opportunity to start a program of urban and technological innovation. (Figs. 7.9 and 7.10)

Fig. 7.9 Some prefigurations and visions for the post-Expo period, in the designs by Carlo Ratti and LAND for Lendlease hypothesis of transforming the area

Fig. 7.10 Some prefigurations and visions for the post-Expo period, in the designs by Carlo Ratti and LAND for Lendlease hypothesis of transforming the area

7.3   Conclusions

The complex horizon structured in these pages tries to demonstrate possibilities, implications and impacts of a deep discussion focused on architectural and landscape quality applied to mega-projects. The megaproject analyzed seems as a very topical case-study: a special and dense process highlights and stresses some needs:

- to read the phenomena at different scales of interactions (from the geographical one of territory to the focused one of architecture);
- to make a synthesis of multiple knowledges (among natural and earth sciences, social and economic disciplines);
- to test decision-making analytical methodologies integrating the singular inputs of the various actors involved in the processes.

The Expo and post-Expo process demonstrated that in megaprojects developments (and mainly in the ones connected with Great Events), the specific topic of their identity and the special occasion is usually overpassing a research towards a spatial quality [11]. But it's exactly the spatial quality the condition because such a large scale and strong transformations can impact in a positive way [12].

We see the urgency to define a clear method of reading and evaluating complex projects, characterized by the participation of various stakeholders, not separating it, as it often happens, from the landscape, urban and architectural design elements and components. Our cultural background is characterized by the strong conviction that they need to be supporting factors, and not secondary ones, in the processes decision-making about territorial transformations.

Our planning and building activity, as well as the political agenda, needs to work on the progressive sharpening of the definition of a multi-scalar and multi-disciplinary approach to the architectural, urban and landscape actions, intended as an integrated project able to measure itself with the transformation processes in an adaptive way, facing decision-making and management operations in an integrated manner.

Being able to work with this approach on an Italian case study is element of further interest as far as the fragmented social, economic and political Italian reality tends to make these large investments, capable of profoundly modifying the environmental and architectural settings of entire assets, very difficult.

As a brief of these essay, we can outline a possible method for a process evaluation matrix aimed at stressing the qualitative aspects of the project, to be added to the quantitative assessments. This matrix can spring from the very topical case study proposed and discussed and thus could become a useful tool for technical and political discussions in analogous situations. But, as well, this matrix could be assumed and applied to different case studies, also of lower scale of transformations and lighter impacts, in wide regeneration and reuse processes, on national and international scales, every time balancing different weights for the different relevant factors.

1. The preliminary projects—or the application dossiers, in the case of Great Events—must contain specific aspects of in-depth analysis with respect to spatial quality issues. These insights cannot be limited to mere declarations of intent or slogans but must instead focus on aspects of a compositional nature;

2. The theme of the large scale, or rather the integration of megaprojects within the territorial order cannot be limited to aspects of an infrastructural nature such as roads and railways. Instead, it must be an opportunity for a wide discussion on ecological and environmental assets: relationship between built and free areas, specific destinations of open areas, waterways, cultivated fields. The environmental footprint of the great transformation must emerge at the various scales of interaction and must become a necessary element of reference and direction, not only as a final verification factor of the non-negativity of the transformation itself;

3. With even greater intensity in those transformations including a strong integration of the natural elements, the planning and the design of megaprojects must develop a specific attention to the time factor. A condition capable of influencing the proposed transformations negatively and/or positively. In some of these, which develop over the years, often some works—normally considered completion, such as planting trees and plants—will have to be programmed as the

first operation, in order to allow the development of essences in the time of construction becoming from the first years of life of the new spatial configuration a determining factor of quality;

4. Strong functional integration is obviously a factor of quality as well as diversity. And it becomes the occasion to build pieces of city or territory that can effectively respond to the functional needs of the contemporary world. However, this integration cannot become synonymous with indiscriminate indifference and openness to any interested economic stakeholder. The establishment of a hotel or a shopping center, a housing district or a school cannot be linked only to the actual request for this function, perhaps to works in progress. The physical and spatial impacts of the different functions are different and require different strategies and actions. The post-Expo case is quite enlightening in this sense, and not in a completely positive sense;

5. The paradox of many megaprojects is that, in the face of the involvement of many stakeholders (including numerous public institutions), the level of sharing of choices, of participation in them, of communication and dissemination of ongoing actions often seems smoky and little effective. The very dimension of involvement must be an opportunity for a wide sharing of choices. Which does not mean mere repetition of often ineffective participatory tools, but rather definition and comparison on a precise mission;

6. The local realization of the works, both in terms of buildings and open spaces, should be the result of extensive competitions and call paths, capable of comparing—within a clear and defined framework—the best solutions: the ones that are able to give concreteness and feasibility to the large scale framework with greater effectiveness and quality. It seems culturally important to point out that, in this framework, the often widespread conviction that the competition—typically a project—is linked exclusively to the aesthetic dimension is overcome. The competition can instead be the occasion for a broad comparison on substantial issues and not, only, of form;

7. While preferring, also for utilitarian reasons and for immediate operation, the new construction, megaprojects can—and must probably—also develop in the redevelopment, regeneration and re-use of the existing, both in terms of buildings and, more generally, of spaces. In a future characterized by instances of control of settlement and resource saving dynamics, this will become a front of commitment and significant development. It is a discourse that also involves the legislative level that will have to be able to make building on the built a viable alternative and not, as often happens, an option that is not sustainable from an economic point of view;

8. The major construction sites—which correspond with important economic resources—must be the main testing ground for architecture and landscape design. It is exactly in these unique and special processes that the constructive culture can grow by experimenting with innovative ideas, testing solutions that, even if expensive in the first instance, can progressively enter into the skills of professionals and companies. It has always been in the DNA of the "big projects" the proposal—also iconic and provocative in some ways (just think about

the Crystal Palace by Joseph Paxton or the Opera House by Jorn Utzon)—of the surprising and unexpected technique or form. The megaprojects today are in a position to play that role;

9. What often turns out to be a limitation of interventions such as megaprojects—that is a self-referentiality, even at a physical and spatial level—must be by-passed in favor of a more marked integration in the body of the city and the territory. The enclaves prove to be short-sighted choices in terms of future development and appropriation by the citizens/users of a place. The first factor of hospitality is configuring a series of connections capable of constructing a condition of marked permeability between places and spaces;

10. Functional integration is also not taken for granted, when we are facing with very strong top-down processes, as it happens in large interventions of infras-tructural nature and extraordinarily strategic (as is the case with Great Events). Instead an integration at different scales of interaction becomes a necessary condition precisely in order to seize the opportunities posed by the transfor-mative processes with significant resources involved: an integration capable also of developing in small triggering progressive processes of overlapping and contamination of possible uses that it can also mean diversity of forms.

References

1. Di Vita S, Morandi C (2018) Mega-events and legacies in post-metropolitan spaces. Expos and Urban Agendas. Palgrave Macmillan, London-New York-Shanghai
2. Guala C (2015) Mega eventi: immagini e legacy dalle Olimpiadi alle Expo. Carocci, Roma
3. Viehoff V, Poynter G (2016) Mega-event cities: urban legacies of global sports events. Routledge, New York
4. Susi Botto I, Di Vita S (2015) EXPO: quale legacy per il futuro di Milano. Planum. J Urbanism 31(II/2015)
5. Sda Bocconi (2016) Camera di Commercio di Milano: L'indotto di Expo 2015
6. Multiple Authors (2014) L'Expo est morte. Vive l'Expo!. Il Giornale dell'Architettura 117
7. Flyvbjerg B (2014) What You should know about megaprojects and why: an overview. Oxford University, Oxford
8. Massidda L (2011) Atlante delle grandi esposizioni universali. Storia e geografia del medium espositivo. Franco Angeli, Milano
9. https://www.ilsole24ore.com/art/impresa-e-territori/2018-08-18/nell-area-expo-primi-segnali-milano-futuro-182539.shtml?uuid=AElZRMVF. Last accessed 20 Nov 2019
10. Arup (2017) Post Expo 2015. Scenarios for a 4D transformation. Planum. J Urbanism 34(I/2017):1–10
11. Roche M (2003) Mega-events, time and modernity: on time structures in global society. Time Soc 12(1):99–126
12. Corner J (ed) (1999) Recovering landscape: essays in contemporary landscape architecture. Princeton Architectural Press, New York
13. Sitography about the most actual visions and perspectives for the area redevelopment. http://www.mindmilano.it/. Last accessed 20 Nov 2019
14. https://www.wired.it/economia/business/2017/11/29/expo-milano-arexpo/?refresh_ce=. Last accessed 20 Nov 2019
15. https://www.lendlease.com. Last accessed 20 Nov 2019

16. https://www.carloratti.com. Last accessed 20 Nov 2019
17. https://www.arexpo.it. Last accessed 20 Nov 2019
18. https://www.landsrl.com. Last accessed 20 Nov 2019
19. https://www.milanocittastato.it. Last accessed 20 Nov 2019

Chapter 8
Megaprojects Contracts: Between International Practices and Italian Law

Francesco Zecchin

Abstract While megaprojects are widely investigated from an economic point of view, there are few studies that analyse them from a legal perspective. The purpose of this paper is to begin to fill this gap, delving into the role that contracts play in the implementation of megaprojects. Given the absence of any international legal framework of reference, the most commonly used contractual models (FIDIC and NEC) will be taken into consideration. After examining the main clauses of such models, their compatibility with the Italian legal system will be explored.

Keywords Megaproject contracts · Duty to performance · Breach of contract · FIDIC models · NEC · Good faith · Delay · Bonds · Defects liability · Dispute · ADR · Arbitration

8.1 The Iron Law of Megaprojects and the Potential of the Legal Instrument

The role of contracts in megaprojects, often also called large engineering projects [46], is almost unexplored by the doctrine of civil law [39]. After all, there is no legal system that has an ad hoc discipline. On the one hand, this prompts the legal practitioner to contend with the phenomenon from a mostly practical perspective, and, on the other hand, it makes the attempt of a general reconstruction particularly complex.

This is all the more so since megaprojects "are not just magnified versions of smaller projects," but "a completely different breed of project in terms of their level of aspiration, lead times, complexity, and stakeholders" [30]. Thus, even at a legal level, things are rather complicated and cannot be tackled by drawing on conventional tools [39]. From the Italian perspective, one also has immediate evidence of this when looking only at the text of the law: the megaprojects contracts must comply with a tangled mass of rules that come from the Civil Code, special laws, and given the

F. Zecchin (✉)
Università Cattolica del Sacro Cuore, via Emiliana Parmense, 84, 29122 Piacenza (PC), Italy
e-mail: francesco.zecchin@unicatt.it

© The Author(s), under exclusive license to Springer Nature Switzerland AG 2020   107
E. Favari and F. Cantoni (eds.), *Megaproject Management*,
PoliMI SpringerBriefs,
https://doi.org/10.1007/978-3-030-39354-0_8

involvement of a public administration in most cases (unlike what always happens more often in other countries [42]), a series of rules of a public nature [8].

As a result, the contractual phase is perceived most of the time as a necessary step for the realization of the megaproject, from which little or nothing can be derived in terms of the chances of success [34]. Indeed, on the part of those observers accustomed to systems such as that in the United States, where the freedom of contract prevails over almost everything [51], it is often believed that the legal instrument, with all its pitfalls, actually represents one of the elements that make up the famous "iron law of megaprojects" [27]. It is normally summarized in these terms: "Over budget, over time, under benefits, over and over again" [29], and it is the fate of roughly ninety percent of megaprojects [31].

Now, the contract can certainly translate into legally acceptable terms a subject matter conceived based on other perspectives, and this often requires changes. This is even more so in the case of megaprojects, which usually involve public interests that the law protects through a series of mandatory rules. This "ordering function" [58] is, however, one of the essential characteristics of law. With particular reference to contracts, it is even the Italian Constitution itself that requires the legislator to determine the limits on the freedom of private economic enterprise for harmonizing its exercise with social utility, with respect for safety, liberty, and human dignity (Article 41, Italian Constitution). This is reflected in the autonomy of negotiations [45].

This does not mean that in Italy the contract is conceived as a means to hindering business activity. On the contrary, although within the boundaries established by the law, the contract is the main instrument that favours the exercise of business activity and its success [10, 63]. Article 41 of the Italian Constitution solemnly confirms it, starting by clearly stating that "Private economic enterprise is free. "And with specific reference to contracts, the second paragraph of Article 1322 of the Italian Civil Code repeats it, stating that "the parties can also consummate contracts that do not belong to the types subject to particular rules, provided they are aimed at achieving interests worthy of protection according to the legal system".

This balance is characteristic of most of the modern legal systems in Continental Europe [50]. In the aftermath of nineteenth-century liberalism, Continental European countries deemed it necessary to codify contractual autonomy within certain legal boundaries, including for the purpose of protecting precisely the freedom of economic enterprise itself [52, 59]. One only needs to consider what it would be like if, for example, there were no rules governing competition in the market or protecting the weak contractor [2]: a freedom only on paper, which is annulled de facto by the monopolist or by the contractually stronger party. It was indeed for good reason that Roman law of centuries ago provided that the structure of interests sought by the parties was not to have been contradictory to the regulatory framework of the era [68].

From this point of view it has been demonstrated that it is not so much the legal instrument itself that contributes to causing the delay or even the failure of megaprojects, but rather the way in which it is handled [67]. Contracts, if well used, represent a resource to govern the risks inherent in megaprojects [40]. They are capable of working across all three aspects that are traditionally identified to assess their level of feasibility: cost, time, and quality [23].

Not surprisingly, in the most recent studies, the contractual aspect is beginning to take on a more and more prominent role [9, 21, 39]. Also because today, as never before, "megaprojects as a delivery model for public and private ventures have never been more in demand, and the size and frequency of megaprojects have never been larger" [30].

8.2   The International Models: FIDIC and the New Engineering Contracts (NEC)

Based on the conviction that the contractual phase does not necessarily represent a complicating factor, but rather a moment for classifying megaprojects, some professional associations have prepared a series of contractual models to be used in the execution of megaprojects. The idea is to make up for the absence of uniform private-law agreements between States [32] and to provide international operators in the sector with a common basis for the various contracts functional to the creation of the megaproject (engineering, procurement,…) [25]. In other words, the legal instrument is no longer an obstacle, but a resource [13].

We can interpret, for instance, the activity carried out by the Geneva-based Fédération Internationale des Ingénieurs-Conseils (FIDIC) in this regard. Following the publications begun in 1957 (Conditions of Contract (International) for Works of Civil Engineering Construction), the federation set up its Contracts Committee. The committee's purpose is to draft "standard forms of contract for use on national and international construction projects"[1] and to update them "to achieve increased clarity, transparency, and certainty, which should lead to fewer disputes and more successful projects" [36]. They are designed especially for financed projects and are grouped into different volumes, divided by situation type. For example, the Red Book is dedicated to suppositions of construction where the work to be carried out is mainly planned by the client; the Yellow Book is addressed to the cases of electrical and mechanical plant and design-build engineering works, and in this case, it is the contractor who draws up the project in light of the employer's indications; the Silver Book, on the contrary, is designed for turnkey construction contracts, where the responsibility for the realization of the project is entirely in the hands of the contractor, who, as the terms implies, will supply a "turn-key" product.

The London-based Institution of Civil Engineers (ICE) has moved in the same direction. Its New Engineering Contracts (NEC) have the purpose of "enabling any project to be delivered on time, within budget and to the highest standards".[2] The first models date back to 1993, and they are now in their fourth edition (2017). They are particularly suitable for cases of infrastructure and utility work, while appearing

---

[1] http://FIDIC.org/bookshop.

[2] Evolving to be the world's favourite procurement suite, NEC, London, 2017.

less practical for projects not requiring a particularly complex organization, such as the construction of buildings [21].

The value of these models in terms of the greater chances for success of the megaprojects is now unanimously recognized. Not surprisingly, they are suggested as Standard Bidding Documents by the major international lenders, such as the World Bank [17] and the European Commission [9], and they are used by many states, such as the United Kingdom, China and South Africa, for the construction of large public works [8].

8.3 FIDIC's and NEC Models' General Principles and Italian Law: (a) Good Faith

General contractors in Italy are also beginning to consider FIDIC and NEC models, especially due to the fact that foreign companies are increasingly involved in the creation of Italian megaprojects. For example, when the employer is a public entity that intends to turn to international contractors, the general terms and conditions and technical specifications are increasingly prepared on the basis of the FIDIC or NEC models [8]. Indeed, in view of differing experiences and the absence of international conventions signed between Italy and other states, the FIDIC and NEC models represent a good starting point for entering into negotiations.

However, the fact remains that, if the project is to be carried out in Italy and especially when it is a public work, Italian law will often be indicated in the contract as applicable law [62]. But even if this were not the case, that is, if the law identified by the parties were that of another State or even there were no choice, the contract would have the closest connection with the Italian legal system. Therefore, the legal criterion of applicable law being that of the state of the contractor's registered office (Article 4, Paragraph 2, Rome I Regulation) would be replaced by that of the location where the work needs to be carried out [60], i.e. Italy. Italian law, therefore, would nonetheless come into play, at least as regards its overriding mandatory provisions [8].

The same reference to Italian law in relation to internationally binding rules would be essential even in the case where the megaproject was needed to be carried out by an Italian company abroad. Again, the FIDIC or NEC models are very frequently used for the bidding process with regard to the project to be realized, and the applicable law is that of the state where the activity must be carried out. However, it is not possible to rule out the consideration of Italy's overriding mandatory provisions pursuant to Article 9, Rome I Regulation.

It is from these standpoints that it becomes necessary to ascertain the impact on the Italian legal system of the solutions identified by FIDIC and ICE. This is all the more important since such models still remain strongly linked to common law systems [9, 54], which do not make a strong distinction between public procurement and private procurement [61] and which, in any case, follow very different principles from those of the civil law systems (such as Italy's) with regard to many contractual aspects.

Considering the models proposed are now very numerous and include a great amount of detail, carrying out a clause-by-clause analysis within a single research paper would be unrealistic [61]. What we can do here, instead, is to analyse some of the general principles underlying the contract models prepared by the FIDIC and the ICE and then compare such principles with the basic ones governing contracts in the Italian legal system. Moreover, this is the direction of research that has also been followed with reference to other systems [9] and appears to be suggested by the associations themselves. In 2019, FIDIC published a volume (The Fidic Golden Principles), whose objective is to identify "which contractual principles of each form of FIDIC contract considers to be inviolable and sacrosanct". The ICE distinguishes the core clauses from the others, which are, in turn, divided according to their substance: main option clauses, secondary option clauses (known as X-clauses), additional conditions of contract (known as Z-clauses), and finally, the new secondary option, X10, specifically to support the use of Building Information Modeling (BIM).

An appropriate starting point is the general issue of good faith, given its relevance in every phase of a megaproject's contracts. Both the NEC and FIDIC models share an increased emphasis on "mutual trust and co-operation" [21]. However, while the NEC contracts expressly propose the adoption of a clause introducing good faith as the general rule in the relations between the parties (10.2), there is no such solution found in the FIDIC contracts.

This obviously does not mean that the parties who adopt one of the FIDIC models have an incentive for behaving improperly. Nevertheless, the risk that the victim of illegal conduct remains devoid of protection will change according to the law that is applicable to the contract.

In actual fact, many FIDIC clauses are based on good faith. Clause 18.2 provides a striking example. According to this clause, the Contractor must give Notice, in the proper form, within 14 days of becoming aware, or of the date when it should have become aware, of the exceptional event. However, there are clauses that seem to leave aside the notion of good faith. One example is Clause 15, which is modelled on the English experience [18]. It deals with the termination of the contract for non-fulfilment and, in some cases, it legitimizes the remedy without subordinating it to a non-performance of no small value [61]. The consequence is that if the good faith does not act as a basis for the contractual relationship (as would be the case if the applicable law were, for instance, English law [44]), a party could free himself from the contractual bond by misleadingly exploiting even minor defaults, and there will be no way to prevent this outcome.

Some problems may also arise when the specific case does not include a clause that explicitly regulates good faith. In these cases, with the changing of the law applicable to the contract, the conduct may or may not be a source of liability.

For example, in England and Wales, case law has repeatedly made clear that there is no general doctrine of good faith in contract law [35]. This means that in the matter of the misleading invocation of a clause of exemption from liability, even if such invocation is established by the contract, the judge has no instruments to prevent the counterparty's improper behaviour [18].

Things are completely different when the applicable law is that of a legal system which recognizes the principle of good faith among the cardinal principles governing contracts. Among these, German law stands out, but there is also Italian law, with Article 1375 of the Italian Civil Code establishing in general that the parties to the execution of a contract must behave properly. So, in the example given immediately above, the misleading invocation of the liability exemption clause would probably be considered illegal. A request to terminate a contract arbitrarily based on a minor breach would probably have the same outcome. And this is the case even if the parties, in terms that are somewhat paradoxical, were to have expressly established that the principle of good faith is not valid between them, since such exception cannot be made in the Italian system [55].

Moreover, if the Italian law were to be the applicable law, the rule of good faith would also extend to the negotiation phase. As in other European jurisdictions, the Italian Civil Code establishes that "the parties must act in good faith in the negotiation and in the formation of the contract" (Article 1337).

This means, for example, that if a party becomes aware of confidential information before the signing of the contract, such information cannot be disclosed. But in the case of megaprojects, nondisclosure agreements are often signed in advance to address this subject.

A much more problematic assumption is that of the negotiations being interrupted when a party is already convinced that the contract will be signed. Here, it is a question of verifying whether who withdrawing its willingness to conclude the contract has behaved improperly, giving rise to the counterparty's expectation of a scenario opposite to what happened [6]. If this were the case, it would be interpreted by Italian law as a violation of good faith, and the party deemed responsible would be ordered to pay damages [19]. Moreover, such interpretation would no longer be based on the rules of unlawful acts, but on contractual liability rules [5].

With regard to this issue there is also a certain difference with the English common law, where a party is not liable for the breakdown of the negotiations even if it were to have made the other party believe that it would have concluded the contract [18]. In actual fact, a different rule is established for procurement and engineering contracts, which are governed by a legislative procedure. In such case, if the specifications have been delivered and the client refuses to sign the contract, after having behaved in a way that made the counterparty reasonably assume that the contract would have been signed, the English courts have sometimes recognized the right to compensation for damages.[3] However, even in Italy, some of the most famous cases of liability for the unjustified breakdown of negotiations have concerned public contracts that were not signed even though the adjudication procedures were at an advanced stage [38]: for example, cases where only the public control authority's approval of the contract was missing [5].

---

[3] *Blackpool & Fidley Aero Club LTD v. Blackpool Borough Council* (1990) 1 W.L.R. 1195 C.A.

8.4 (b) Delay and Liability

The issue of the timing of fulfilment is one of the most delicate in megaprojects contracts. Suffice it to say that so-called "critical path diagrams" have been developed for this sector to illustrate the schedule of the planned and contractually agreed execution times for each segment—minor or major—in which the works are divided [14]. With regard to the distribution of the risk of delays in contract fulfilment, the FIDIC model and NEC provide for different rules as well.

In the early versions, the FIDIC models referred to "force majeure", defining it as "an event which is beyond Party's control, which such Party could not reasonably have provided against before entering into the contract, which, having arisen, such Party could not reasonably have avoided or overcome; and which is not substantially attributable to the other Party". If all these conditions occur, the debtor is not liable for the delay and, on the contrary, is entitled to an extension of time and a payment for any such cost (clause 18.4).

In the 2017 version, a change was made to the heading of the clause dedicated to this aspect, n. 18. It is no longer "force majeure", but "exceptional events". This linguistic change shifts the FIDIC models further away from the lexicon of civil law [41] and removes any doubt about the fact that if this clause is adopted, the debtor cannot be held not liable by only demonstrating that he is not responsible for the impossibility of the performance.

Instead, the debtor must specifically prove the occurrence outside his sphere of control that made fulfilment impossible. The event cannot remain unknown, otherwise the debtor will be liable. For example, in the event of the failure to execute works due to tampering with the machinery necessary for the realization of the works, it is not enough for the debtor to prove that he did everything he had to prevent this possibility. Firstly, he must prove that a third party tampered with the machinery without his being able to prevent it. Secondly, there must be proof that this impediment could not have been contemplated at the time of the signing of the contract, otherwise it is assumed that the debtor took on the risk of the impediment. The arrival of a cyclone, therefore, cannot be predicted only if the work is to be carried out during a period of the year in which cyclones never occur. Thirdly, for an exemption from liability, it is necessary that the event, having arisen, was not reasonably avoidable or surmountable. Therefore, in an earthquake zone, the destruction of the partially constructed project because it has not been made safe after the issuance of a tsunami warning is not covered by the clause. Finally, the event must have occurred without the fault of either party. Therefore, the debtor is not exonerated if he is late in completing the project and an earthquake occurs in a non-seismic area causing damage that would not have occurred if the project had been completed on time.

The criterion underlying the FIDIC models seems to draw inspiration from the economic analysis of law: the risk is allocated to the party who can best manage it and can sustain its consequences [9, 14, 26, 40, 56]. The debtor will be liable for all events that fall within his sphere of control—what in Italy is called "typical risk" [65]. This means, for instance, that the breakage of essential machinery for the realization of the project, even when it is unpredictable and unavoidable, cannot exempt the debtor from liability. The same applies to the actions of those for whom the debtor is liable, especially those that he commissions, such as subcontractors, to perform the works.

In other words, it is an objective responsibility, which does not consider the conduct of the debtor and is based on the traceability of the event among the events inherent to its activity. Aside from these assumptions, that is, when the impossibility of the performance is due to a cause extraneous to risks controllable by the debtor, the weight of the damage, if nothing is provided and the law applicable to the contract does not establish otherwise, will be borne by the creditor, who can address such risk by entering into an insurance policy [66].

Another idea underlying the rule of force majeure is that it would act as a deterrent to fulfilment [39]. From this point of view, however, there is the danger of introducing a counterproductive mechanism, because it can make fulfilment even more unlikely [39]. Indeed, the obligation of compensation, which is normally based on a penalty much higher than the actual damage, very often does not have any function as a deterrent. On the contrary, it ends up further aggravating what is already, in most cases, a difficult situation for the debtor, and it is precisely the reason for the delay [30].

Not surprisingly, force majeure is not even mentioned in the NEC and clause 60.1 (19) provides that the debtor is exempt from liability when he proves "an event which stops the Contractor from completing the whole of the works or stops the Contractor from completing the whole of the works by the date shown on the Accepted Programme and which neither Party could prevent, an experienced contractor would have judged at the Contractor Date to have such a small chance of occurring that it would have been unreasonable for him have allowed for it and is not one of the other compensation events stated in the contract". This type of clause is less strict than that identified in the FIDICs models. Indeed, the risk of the debtor's activity is not at its expense. To free himself from liability, the debtor only needs to prove an unforeseeable event, for an experienced debtor too, which makes contractual fulfilment impossible and has not been specifically considered in the contract. One example might be the sudden breakage of new machinery, or a hurricane in an area where hurricanes rarely occur. The concept behind this model seems to be that of a more equitable distribution of risk among the parties based on a relationship of reliable collaboration and with a view toward realizing the project as quickly as possible, rather than the normal practice of assigning the debtor the responsibility for the delay and, through a penalty, making it even more difficult compared to what it is already [40].

The solution of the NECs models is not very far from that generally given by the combined provisions of Articles 1218 and 1176 of the Italian Civil Code (and in the case of private procurements, indirectly by Article 1672 of the Italian Civil Code, while in the case of public procurements, by Article 107, Paragraph 5 of Legislative Decree No. 50 of 18 April 2016). According to these provisions, the debtor must be exempt from liability on the basis of a criterion which is less strict than that of the force majeure. It is sufficient for the debtor to prove that he was faced with the impossibility of the performance which could not be surpassed with normal diligence [45]. In other words, the debtor must prove that, despite having taken all the precautionary measures of the bonus pater familias, an unforeseeable event occurred that prevented him for fulfilling the contract. It is a burden of proof very different from the force majeure that is provided by the FIDIC models and Italy's legal system refers to it as a "fortuitous event" [57].

In actual fact, there seem to be no obstacles to which the parties might derogate from the Italian rules by introducing a broader criterion of imputability of the impossibility. This is all the more so since, according to a long-standing jurisprudential orientation, the rule of Article 1218 of the Italian Civil Code would oblige the debtor to prove the fact that he made the fulfilment impossible.[4] Serious doubts can arise that this is so and, in fact, this orientation is not often followed [4], if not in terms of the recognition that it is often de facto easier for the debtor to prove the fortuitous event rather than his diligence [45].

At the same time, however, it has been noted that an aggravated liability is contemplated pursuant to some provisions of the Italian Civil Code in relation to contracts in which the debtor is normally an entrepreneur (Articles 1693, 1787 and 1893 of the Civil Code). This gives rise to the theory that the rule referenced in Articles 1218 and 1176 of the Civil Code has a subsidiary role in the cases in which the debtor is a person who carries out an organized economic activity. Such rule is applicable only when the law provides for liability that is not extended to the fortuitous event and is contained within the limits of diligence [45]. In other words, in these cases the criterion would be that of allocating the risk to the person who is able to manage it [53].

8.5 (c) Variations

Variations are one of the most important issues for megaprojects contracts, as these contracts have a term for which the emergence of unforeseen events is quite inevitable. On the other hand, flexibility is deemed one of the essential properties for the realization of megaprojects [11] and it has a form of privileged expression within the contractual instrument [28].

Not surprisingly, both the FIDIC and the NEC models propose the insertion of a hardship clause in relation to variations. Both models refer to some extent to the traditional distinction between variations that are only improving the project and variations that are necessary for its completion [41]. But they use such different terminologies and rules that they cannot be considered as overlapping.

The FIDIC models deal with variations in clause 13 and they talk about variations to refer to all the changes that are required for various reasons by the parties and subject to an Engineer evaluation. They refer to the concept of adjustments for changes in relation to the modifications due to changes in legislation or costs. This

---

[4]See Cass. 22 December 1978, n. 6141, in *Giur. it.*, 1979, I, 1, p. 953s.

clause needs to be read in conjunction with clause 12, entitled Measurement and Evaluation, and clause 20.1, dedicated to the Contractor's Claims.

The NEC models deal with the cases of variations and claims together, with the latter defined as requests made by the debtor for the incremental costs involved in executing the project. The clause 14.3 is dedicated to the employer's instructions, while the numbers 60–66 concern the compensation events, defined as those events which are not the fault of the contractor, but that not necessarily the fault of anyone else; they could be fault-neutral, such as weather conditions or site conditions.

Both models provide for the employer's broad discretion, albeit based on somewhat different criteria, in the requests for changes and an obligatory renegotiation procedure to be pursued in the event of variations not related to the will of the parties [9, 61]. Furthermore, they identify a person assigned with the task of mediating between the parties: the Engineer for the FIDICs and the Project Manager for the NECs. These clauses are based on the criterion of economic efficiency. Their purpose is to get to the solution of the problem as quickly as possible and without disputes, protecting both parties' interests to the maximum extent possible [7].

With regard to the issue of variations, it is necessary to point out the gap with respect to the Italian legal system, in which similar rules are provided for the procurement contract only. In these cases, the Civil Code establishes a real renegotiation obligation when it comes to changes necessary to the project and, if the parties do not agree, the task of determining the variations and the correlated price changes is delegated to a judge (Article 1660, Paragraph 1 of the Civil Code). However, if these variations exceed a certain limit, the second and third paragraphs of the aforementioned article provide for the right of withdrawal by the customer and the contractor, subject to fair compensation to the latter. Instead, with regard to the changes requested by the client, Article 1661 of the Civil Code recognizes this right only if their amount does not exceed 16.6% of the agreed price, after which, or even in the event of significant changes, the contractor has the right to execute the project as it had been agreed. Finally, Article 1664 of the Civil Code provides that in cases of an increase beyond a certain limit (10%) of the cost of execution, it is possible to request a revision of the price. Furthermore, if there are execution difficulties during the project as a result of geological, water-related, and similar causes not foreseen by the parties, and these difficulties make the contractor's service considerably more expensive, the contractor has the right to fair compensation. Fairly similar rules, at least in terms of their *ratio*, are provided by Article 106 of Legislative Decree No. 50 of 18 April 2016 [49]. Besides, in the execution phase, the public interest ceases to exist, and the relationship is essentially treated as a contract between private individuals [37].

Instead, outside the sphere of procurement contracts, the Italian legal system does not have a general rule that requires a party to comply with the new indications received from the other party [59] or obligations to renegotiate in case of contingencies [20]. Unless the parties have chosen, as the applicable law, the law of a state other than Italy which follows principles like these, two options remain.

If the parties explicitly want to establish a different rule, they will have to expressly agree with a specific clause that introduces a ius variandi for voluntary changes and a hardship clause for contingencies. In this case, however, it is necessary to ensure that such agreements are not the result of an abuse of the greater bargaining power that derives from the economic dependence of one party on the other. Such a situation could put the validity of the contract at risk [1]. Moreover, the outcome would be the same even if the applicable law were not Italian law, given that the regulations normally protecting the weak contractor within the Italian legal system are binding internationally [64]. In this regard, it is necessary both to be very prudent in the use of the FIDIC or NEC clauses, which vest very broad power with the client [61], and to recognize that a major shift away from the rules established by the Civil Code for tenders could be risky.

On the other hand, should the parties not regulate these aspects through ad hoc clauses, it will be necessary to rely on the debtor's receptiveness to the variations determined by the will of the parties. Except for what is due in accordance with good faith pursuant to Article 1375 of the Italian Civil Code, the debtor has the right to perform the service as it had been agreed [59]. Instead, in the case of external events that do not cause the contract obligations to become excessively onerous pursuant to Article 1467 of the Italian Civil Code, which gives the right to avoid termination by modifying the terms of the contract fairly (Paragraph 3), the weight will again be distributed according to the rule of Article 1218 of the Italian Civil Code or, if the theory of business risk is adopted, the criterion of force majeure [33].

An alternative remains as a backup, but this does not offer sufficient margins of certainty and, therefore, it leads to scenarios that are difficult to predict. With the tender rules not being deemed exceptional pursuant to Article 14 of the preliminary provisions, it could be possible assume a similar application of such rules for the other contracts that refer to the megaproject and that respond to the same ratio, such as, for example, the engineering contract [47].

8.6 (d) Bonds and Defects Liability

Guarantees represent another delicate issue in megaprojects, as they are used for protecting the huge investments that the client needs to make for the megaproject realization [54]. Broadly understood, guarantees cover not only the phase following the completion of the project, but also the phases of adjudication and execution [61]. With reference to adjudication, bind or tender bonds have been more and more frequently used for the purpose of compensating damages that the client might suffer if the contractor were not to enter into the contract after having been adjudicated the winner [8]. In relation to the execution phase, advanced payment bonds serve to guarantee the refund of the advance payment given for the start-up of the works, while performance bonds are used to guarantee the proper fulfilment of the project [48].

Through various clauses, both the FIDIC and the NEC models propose these types of bonds [54]. Italian legilsator has moved in the same direction for the public contracts, through Articles 93 and 103 of Legislative Decree No. 50 of 18 April 2016.

Such rules provide for a series of guarantees that seem, at least in part, similar to those used in the FIDIC and NEC models and in international practice [48]. At the same time, however, these provisions of Italian law provide for a series of limitations, which have been somewhat diminished through recent reform [54], but which still entail points of difference with the FIDIC and NEC models. One example is the fact that the guarantees must expressly provide for the exclusion of the beneficium excussionis and they are by simple demand (Article 93, Paragraph 4, and Article 103, Paragraph 4 of Legislative Decree 50/16). These are mandatory rules, which must therefore be duly considered so as not to invalidate the clauses.

Nothing, however, is established by Italian law with reference to relationships between private parties, not even in the procurement contract. In actual fact, there seem to be no obvious reasons that would prevent, at least in the abstract, the incorporation in megaprojects contracts of the clauses set out in FIDICs and NECs on the subject of bonds. Even more considering that the adoption of first-demand guarantees is not recommended, for the purpose of avoiding abuses in most cases, and, in our legal system, such guarantees have a special role in relationships between private parties [16]. It is, instead, recommended that the creditor will have specifically indicate which is the defaulted obligation [54]. This is, however, once again a matter of avoiding that these clauses, especially those that establish a performance bond, will result from the abuse of power given by the dominant position of one party over the other.

Maintenance bonds or warranty bonds are the guarantees relating to the defects in the project work executed. It is commonplace that the contractor is liable to remedy any defects on completion of the works. The defects must be reported before the end of a specified period, called the defects notification period. Both the FIDIC (clause 11) and the NEC (clause 43) follow this rule and provide for a defect correction period within which the contractor must solve the problem and, absent thereof, the possibility of the employer handling it and then charging back the cost to the contractor.

A first problematic aspect is that these models draw no line to distinguish between the defects for which the contractor is responsible because they are really attributable to him and the defects for which the contractor is asked to be responsible for, even if they are not attributable to him. Even though many parties, including the employer, contribute to the realization of megaprojects, the outcome is that the contractor is always responsible for the defects. From this point of view, the Italian legal system, at least in cases of private procurement, takes particular care to the contractor. Indeed, there are prospects allowing the possibility that the contractor can be released from the guarantee, arguing that the defect stems from a decision made and imposed by the client [15].

Furthermore, the FIDIC and NEC models do not distinguish latent defects from patent defects, so, even for the latter, one can assume that the guarantee and the notification period are valid. This is not the case in the Italian legal system, which in the field of private procurement still considers it more balanced to establish that if the work was accepted and the defects were recognizable, the guarantee is not enforceable. And even in public procurement, the recognizable defects must be reported before the test certificate becomes definitive. These are rules to protect the customer, which therefore must not be neglected in drafting of the contract.

8.7 (e) Disputes and Arbitration

Finally, it is worth mentioning the issue of resolving possible disputes between the parties. In general, both models are aimed at avoiding court proceedings, to the extent possible, as such proceedings can lead to a suspension of the works and to the risk of the megaprojects not ever being completed. The models, therefore, provide for series of intermediate steps in preparation for the actual adjudication, which is usually in the form of an award issued by an international arbitration panel [62].

In this regard, the FIDIC's 1994 edition identified the engineer as the third party in charge of resolving any disputes that arose between the parties. The mechanism provided for three phases: engineer's decision, conciliation attempt in case of disagreement about the engineer's decision, and, finally, arbitration. The model vested significant power with the engineer mainly because of his in-depth knowledge of the project. However, this did not succeed in counterbalancing the issue of the engineer partiality. In fact, the engineer is closely connected to the client, and it is accordingly difficult for it to be deemed a super partes entity [12]. Italian case law on this subject has also shown some inconsistencies, with one ruling stating that the engineer's decision could not be qualified as an arbitration award,[5] and another ruling instead indicating that it represented an informal award pursuant to the 1958 New York Convention.[6]

For this reason, the FIDIC models have replaced the engineer, which remains responsible for evaluating only the claims of the contractor (20.1), with a Dispute Adjudication Board (20.2). In the NEC model, the same undertaking is delegated to a Dispute Avoidance Board (W3). The boards normally consist of one or three industry professionals, who are independent, follow the project from the beginning, and, to the extent possible, even have a remit to pro-actively identify issues before they become disputes. Their decisions are not comparable to those of an arbitration panel, since the purpose is not just to decide who is right and who is wrong in accordance with the law, but to find, as quickly as possible, a solution that is satisfactory to both parties and allows the continuation of the megaproject [24]. This aspect can also be seen from a merely linguistic perspective. For example, the NECs do not cite "disputes" to the board, but rather "potential disputes".

The solutions fall within the framework of Alternative Dispute Resolution (ADR) and, in more specific terms, represent cases of mediation [43]. The characteristic trait that distinguishes these scenarios from legal proceedings and arbitration in the Italian legal system too is the intention to find a solution satisfactory to both parties, without derogating from the judicial authority [22] and not on the basis of the alternative of

---

[5] App. Roma, 12 December 1994, in www.dejure.it.

[6] App. Roma, 21 July 1997, in www.dejure.it.

right or wrong [3]. In this perspective, with regard to public procurement, the parties will need to consider the rules established by Law No. 55 of 4 June 2019. Through Paragraphs 11, 12, 13 and 14, this law has reintroduced the possibility of setting up a technical advisory board, which is charged with providing assistance for the rapid resolution of disputes of any nature likely to arise during the execution of the contract. The decision in the case of private contracts is less constrained, even though for this assumption, it is also necessary to consider the reformed text of Article 5, Paragraph 5 of Legislative Decree No. 18 of 4 March 2010. Such text does not establish that conciliation clauses of a contractual nature introduce a condition of prosecutability of the action. Nevertheless, if the mediation attempt has not been made, it is limited to providing that the judge or the arbitrator assigns the parties the term of fifteen days for the filing of the request for mediation and sets the next hearing after the expiry of that term. However, it will be possible to demand damage compensation of the person who has failed to perform what was required by the contract [22].

References

1. Albanese A (1999) Abuso di dipendenza economica: nullità del contratto e riequilibrio del rapporto. Europa e diritto privato, pp 1179–1220
2. Albanese A (2008) Contratto mercato responsabilità. Giuffrè, Milano
3. Albanese A (2012) Dalla giurisdizione alla conciliazione. Riflessioni sulla mediazione nelle controversie civili e commerciali. Europa e diritto privato, pp 237–253
4. Albanese A (2014) Il rapporto obbligatorio: profili strutturali e funzionali. Libellula, Tricase
5. Albanese A (2017) La lunga marcia della responsabilità precontrattuale: dalla *culpa in contrahendo* alla violazione degli obblighi di protezione. Europa e diritto privato, pp 1129–1148
6. Albanese A (2018) Responsabilità precontrattuale. In: LE PAROLE DEL DIRITTO. SCRITTI IN ONORE DI CARLO CASTRONOVO, III, Jovene, Napoli, pp 1695–1726
7. Amore G (2007) Appalto e *claim*. Cedam, Padova
8. Balestra L (2016) Le regole applicabili alla formazione del contratto internazionale di appalto: standard internazionali e discipline nazionali. Corriere giuridico, pp 634–647
9. Besaiso H, Fenn P, Emsley M, Wright D (2018) A comparison of the suitability of FIDIC and NEC conditions of contract in Palestine. Eng Construc Architec Manage 25(2):241–254
10. Betti E (1953) Teoria generale delle obbligazioni I. Prolegomeni: funzione economico-sociale dei rapporti di obbligazione, Giuffrè, Milano
11. Bettis RA, Hitt MA (1995) The new competitive landscape. Strateg Manag J 16:7–19
12. Bowcock J (1997) The new supplement to the FIDIC: red book. Int Construc Rev 49–60
13. Bucci S (2012) L'appalto internazionale. In: Patroni Griffi U (eds) MANUALE DI DIRITTO COMMERCIALE INTERNAZIONALE, Giuffrè, Milano, pp 186–191
14. Bunni N (2013) The FIDIC forms of contract, 3rd edn. Blackwell Publishing Ltd., Oxford
15. Cappai F (2011) La natura della garanzia nei vizi dell'appalto. Giuffrè, Milano
16. Cappai F (2015) Il contratto autonomo di garanzia nel commercio internazionale. Rivista di diritto civile II:127–157
17. Cardinale E (2001) Il contratto tipo nel commercio internazionale. In: Patroni Griffi U (ed) MANUALE DI DIRITTO COMMERCIALE INTERNAZIONALE. Milano, Giuffrè, pp 91–107
18. Castronovo C (2001) Principi di diritto europeo dei contratti. Parte I e II, edizione italiana, Giuffrè, Milano
19. Castronovo C (2018) Responsabilità civile. Milano, Giuffrè

20. Cataudella A (2019) I contratti. Parte generale, 5th edn. Giappichelli, Torino
21. Close J (2017) A comparative guide to standard form construction and engineering contracts. Law Brief Publishing, Somerset
22. Curti M (2002) Verso una tipicizzazione della c.d. « clausola di conciliazione contrattuale ». Nuova giurisprudenza civile commentata, I, pp 17–19
23. Dimitriou H, Ward EJ, Wright PG (2012) Mega projects executive summary: lessons for decision makers: an analysis of selected international large-scale transport infrastructure projects. Omega Centre, Barlett School of Planning, University College London, London
24. Dorgan CS (2005) The ICC's new dispute board rules. Int Construc Law Rev 22:142–150
25. Draetta U (1984) Il diritto dei contratti internazionali. La formazione dei contratti, Cedam, Padova
26. Eggleston B (2006) The NEC 3 engineering and construction contract a commentary. Blackwell Science, Oxford
27. Fenn P, Lowe D, Speck C (1997) Conflict and dispute in construction. Construc Manage Econ 15:513–518
28. Floricel S, Miller R (2001) Strategizing for anticipated risks and turbulence in large-scale engineering projects. Int J Project Manage 19:445–455
29. Flyvbjerg B (2011) Over budget, over time, over and over again: managing major projects. In: Morris PWG, Pinto JK, Söderlund J (eds) The Oxford handbook of project management. Oxford University Press, Oxford, pp 321–344
30. Flyvbjerg B (2014) What you should know about megaprojects, and why: an overview. Project Manage Inst 45:6–19
31. Flyvbjerg B (2017) Introduction: the iron law of megaproject management. In: Flyvbjerg B (ed) The Oxford handbook of megaproject management. Oxford University Press, Oxford, pp 1–18
32. Galgano F, Marrella L (2010) Diritto e prassi del commercio internazionale. Cedam, Padova
33. Gallo P (2011) Revisione e rinegoziazione del contratto, Digesto delle discipline privatistiche Sezione civile Aggiornamento, pp 804–821
34. Glenn S (1992) Contracts cause conflicts. In: Fenn P, Gamenson R (eds) Construction conflict: management and resolution. Chapman & Hall, London, pp 128–144
35. Glover G (2017) Introducing NEC4. Int Q 22:1–5
36. Glover G (2018) Some thoughts on how the 2017 FIDIC contract deals with time. Int Q 24:1–12
37. Giannini MS (1988) Diritto amministrativo, I. Giuffrè, Milano
38. Ilacqua A (2009) La responsabilità precontrattuale della pubblica amministrazione. Evoluzioni giurisprudenziali. Giustizia amministrativa, pp 418–435
39. Jobling PE, Smith NJ (2018) Experience of the role of contracts in megaproject execution. Proc Inst Civil Eng Manage Procure Law 171:18–24
40. Jobling PE, Smith NJ, del Rey F (2019) Discussion: experience of the role of contracts in megaproject execution. Proc Inst Civil Eng Manage Procure Law 172:36–37
41. Klee L (2018) International construction contract law. Wiley, Hoboken
42. Marrewijk A (2005) Strategies of cooperation: control and commitment in mega-projects. Management 8:89–104
43. Mazzamuto S, Plaia A (2007) I rimedi. In: Castronovo C, Mazzamuto S (eds) MANUALE DI DIRITTO PRIVATO EUROPEO, II. Giuffrè, Milano, pp 739–814
44. McKendrick E (2007) La buona fede tra *common law* e diritto europeo. In: Castronovo C, Mazzamuto S (eds) MANUALE DI DIRITTO PRIVATO EUROPEO, II. Giuffrè, Milano, pp 715–735
45. Mengoni L (2011) Scritti II. Obbligazioni e negozio. Giuffrè, Milano
46. Miller R, Lessard DR (2019) Evolving strategy: risk management and the shaping of large engineering projects. MIT Sloan Research Paper No. 4639–07. Available at SSRN: https://ssrn.com/abstract=962460 or http://dx.doi.org/10.2139/ssrn.962460. Last accessed 24 Nov 2019
47. Musolino G (2003) L'appalto internazionale. Giuffrè, Milano
48. Musolino G (2006) Responsabilità e garanzie nell'appalto internazionale. Rivista trimestrale degli appalti, pp 2–45

49. Musolino G (2016) L'esecuzione dell'appalto nel nuovo codice dei contratti pubblici. Rivista trimestrale degli appalti, pp 457–502
50. Nicolussi N (2014) Etica del contratto e « Contratti "di durata" per l'esistenza della persona». In: Nogler L, Reifner U (eds) LIFE TIME CONTRACTS: SOCIAL LONG-TERM CONTRACTS IN LABOUR, TENANCY AND CONSUMER CREDIT LAW. Eleven International Publishing, Den Haag, pp 123–167
51. Nicolussi N (2014) *Enhancement* e salute nel rapporto medico paziente. In: Palazzani L (ed) VERSO LA SALUTE PERFETTA. *ENHANCEMENT* TRA BIOETICA E BIODIRITTO. Edizioni Studium, Roma, pp 89–124
52. Nivarra L (2016) Relazione introduttiva. In: Mazzamuto S, Nivarra L (eds) GIURISPRUDENZA PER PRINCIPI E AUTONOMIA PRIVATA. Giappichelli, Torino, pp 3–10
53. Pardolesi R, Mattei U, Monateri PG, Cooter R, Ulen T (1999) Il mercato delle regole. Il Mulino, Bologna
54. Patroni Griffi U (2001) Appalti interni ed internazionali di costruzioni: la disciplina delle garanzie. Riv. comm. int., pp 411–419
55. Piraino F (2015) La buona fede in senso oggettivo. Giappichelli, Torino
56. Potts K (2008) Construction cost management-learning from case studies. Taylor & Francis, Abingdon
57. Realmonte F (1988) Caso fortuito e forza maggiore. Digesto disc. priv. – sez. civ., II, Torino, Utet
58. Romano S (1947) Frammenti di un dizionario giuridico. Giuffrè, Milano
59. Roppo V (2011) Il contratto, 2nd edn. Giuffrè, Milano
60. Rossi EA (2016) Fenomenologia delle esternalizzazioni: il contributo della prassi internazionale ed europea alla disciplina degli appalti privati. Contratto e impresa – Europa, pp 66–98
61. Rubino-Sammartano M (2006) Appalti di opere e contratti di servizi (in diritto privato), 2nd edn. Cedam, Padova
62. Rubino-Sammartano M (2017) Specificità dell'appalto internazionale (e, sotto vari profili, dell'appalto in genere). Rivista trimestrale degli appalti, pp 169–174
63. Sacco R (2005) Contratto, autonomia, mercato. In: Sacco R, De Nova G (eds) IL CONTRATTO, 3rd edn. 1, Giuffrè, Milano, pp. 16–48 (2005)
64. Seatzu F (2016) L'appalto nel diritto internazionale privato dell'Unione europea. In: Luminoso A (eds) CODICE DELL'APPALTO PRIVATO, 2nd edn. Giuffrè, Milano, pp 132–164
65. Trimarchi P (1970) Sul significato economico dei criteri di responsabilità contrattuale. Riv. trim. dir. proc. civ., pp 512–531
66. Trimarchi P (2010) Il contratto: inadempimento e rimedi. Giuffrè, Milano
67. Valentin WS, Vorster FS (2012) Understanding construction project failure in Southern Africa. In: Proceedings of the institution of Civil Engineers—Management, Procurement and Law 165:19–26
68. Zimmermann R (1996) The law of obligations. Roman Foundations of the Civilian Tradition, Clarendon Paperbacks, Oxford

Index

Printed in the United States
By Bookmasters